Lecture Notes in Control and Information Sciences

Volume 458

About this Series

This series aims to report new developments in the fields of control and information sciences—quickly, informally and at a high level. The type of material considered for publication includes:

1. Preliminary drafts of monographs and advanced textbooks
2. Lectures on a new field, or presenting a new angle on a classical field
3. Research reports
4. Reports of meetings, provided they are

   (a) of exceptional interest and
   (b) devoted to a specific topic. The timeliness of subject material is very important.

More information about this series at http://www.springer.com/series/642

Alexander L. Zuyev

Partial Stabilization and Control of Distributed Parameter Systems with Elastic Elements

 Springer

Alexander L. Zuyev
Institute of Applied Mathematics
  and Mechanics
National Academy of Sciences of Ukraine
Donetsk
Ukraine

ISSN 0170-8643   ISSN 1610-7411   (electronic)
ISBN 978-3-319-11531-3   ISBN 978-3-319-11532-0   (eBook)
DOI 10.1007/978-3-319-11532-0

Library of Congress Control Number: 2014950653

Springer Cham Heidelberg New York Dordrecht London

Printed on acid-free paper

Springer is part of Springer Science+Business Media (www.springer.com)

To my mother Lidia,
and to the memory of my father Leonid

Preface

In recent years, problems of stability and control of nonlinear dynamical systems governed by partial differential equations and abstract differential equations have become an intensive field of research. We refer to works by Barbu [1], Coron [2], Curtain and Zwart [3], Fattorini [4], Krstic and Smyshlyaev [5], Lasiecka and Triggiani [6] as basic monographs in this area. The importance of this study is underpinned by the fact that systems of coupled nonlinear ordinary and partial differential equations describe the evolution of flexible structures, such as controlled flexible manipulators, spacecrafts endowed with elastic antennae, solar panels, and tethers. Illustrious examples of the control of models of flexible structures with elastic beams and plates are presented in monographs by Junkins and Kim [7], Komkov [8], Krabs [9], Lagnese [10], Lagnese et al. [11], Luo et al. [12], Meurer [13], Nabiullin [14], Oostveen [15], Sirazetdinov et al. [16]. Despite a vast literature on the control of flexible structures, there is still a gap in the qualitative theory of distributed parameter systems concerning the problems of partial stability and control.

The concept of stability with respect to a part of variables, introduced by Lyapunov [17], has been intensively studied in monographs by Rumyantsev and Oziraner [18], and Vorotnikov [19]. The significance of this study is justified by a well-known fact that a control system with integrals cannot be stabilized in the strong sense, but only the partial stabilization is possible for many physical systems. The majority of publications in the partial stability theory deal with finite-dimensional systems described by ordinary differential equations. The goal of this book is to provide a rigorous treatment of problems related to partial asymptotic stability and controllability for classes of infinite-dimensional mechanical systems with elastic elements. Our study is mostly based on Lyapunov's direct method, and we do not intend to cover here other approaches of the distributed parameter control theory, such as backstepping design, flatness, frequency-domain, or spectral methods. The interested reader can follow through the developments of these approaches in monographs [6, 13, 20–23].

In order to keep the book self-contained, some basic results from the theory of continuous semigroups of operators in Banach spaces are presented in Chap. 1. The rest of this book is organized as follows.

In Chap. 2, the problem of partial asymptotic stability with respect to a continuous functional is considered for a class of abstract multivalued systems on a metric space. A sufficient stability condition is derived by means of Lyapunov's direct method and the invariance principle. For the case of nonlinear continuous semigroups, asymptotic stability is characterized in terms of a differentiable Lyapunov functional. In finite-dimensional spaces, the problem of partial stabilization is considered for differential inclusions and autonomous systems of ordinary differential equations. A result on sufficient condition for partial stabilizability is proved, provided that the open-loop system admits a Lyapunov function with nonpositive lower bounds of time-derivatives. The result proposed is applied to examples of single-axis stabilization of a satellite. Two cases are considered. In the first case, the attitude is controlled by thrust jets, while in the second case the satellite is controlled by means of a pair of flywheels. Explicit expressions of stabilizing feedback laws are given.

A mathematical model of a rotating body with elastic attachments is introduced in Chap. 3. It is shown that the equilibrium of the considered system cannot be made strongly asymptotically stable in the general case. This brings the motivation for considering the problem of partial stabilization with respect to a functional that represents "averaged" oscillations of the system. Such a problem is studied in detail in Sects. 3.1–3.4.

Section 3.5 is concentrated on the modeling of a robotic manipulator with flexible and rigid parts. The dynamics is described by a system of coupled ordinary and partial differential equations, which is transformed into an abstract differential equation in a suitable Hilbert space. A feedback control that ensures strong asymptotic stability of the equilibrium is proposed.

Chapter 4 is focused on the spillover analysis for infinite-dimensional systems with finite-dimensional controls. It is shown that a family of L^2-minimal controls, corresponding to low-frequency modes, can be used to solve approximately the steering problem for the complete system. This control design scheme is applied to the Euler–Bernoulli beam.

A mathematical model of a flexible-link manipulator is derived in Chap. 5 by exploiting the Timoshenko beam theory. The controller design is proposed for Galerkin's approximations with an arbitrary number of elastic modes.

In Chap. 6, we consider a mechanical system consisting of a rigid body with a thin elastic plate. The plate vibrations are governed by the Kirchhoff theory. We derive the equations of motion as a system of ordinary and partial differential equations and consider the angular acceleration of the body as the control. The dynamical equations are transformed to an infinite set of ordinary differential equations with respect to elastic coordinates. It is shown that such a system is neither controllable nor stabilizable in general. A theorem on partial stabilization of the equilibrium by a state feedback law is proved in Sect. 6.2. Then the approximate controllability problem is set out for the dynamics restricted to an invariant manifold. An estimate of the reachable set is proposed by using the approach of Chap. 4, and sufficient conditions for the approximate controllability are obtained as a corollary of this estimate. Simulation results are presented to illustrate the spill-over effect.

Many parts of this book exploit representations of solutions by the Fourier series with elastic coordinates or approximations by Galerkin's method. Such representations appear quite natural from the point of view of analytical mechanics, where generalized coordinates are used to characterize the dominant dynamics of a large-scale system. To justify this paradigm more rigorously, we prove the convergence of Galerkin's method in Appendix A for the Euler–Bernoulli beam model.

The author is grateful to Dr. Victoria Grushkovskaya for the assistance in typesetting some parts of this book.

Donetsk, July 2014   Alexander L. Zuyev

References

1. Barbu, V.: Analysis and Control of Nonlinear Infinite Dimensional Systems. Academic Press, San Diego (1992)
2. Coron, J.-M.: Control and Nonlinearity. AMS, Providence (2007)
3. Curtain, R.F., Zwart, H.: An Introduction to Infinite-Dimensional Linear Systems Theory. Springer-Verlag, New York (1995)
4. Fattorini, H.O.: Infinite Dimensional Optimization and Control Theory. Cambridge University Press, Cambridge (1999)
5. Krstic, M., Smyshlyaev, A.: Boundary Control of PDEs: A Course on Backstepping Design. SIAM (2008)
6. Lasiecka, I., Triggiani, R.: Control Theory for Partial Differential Equations: Continuous and Approximation Theories. 2: Abstract Hyperbolic-like Systems over a Finite Time Horizon. Cambridge University Press, Cambridge (2000)
7. Junkins, J.L., Kim, Y.: Introduction to Dynamics and Control of Flexible Structures. AIAA Education Series. AIAA, Reston, VA (1993)
8. Komkov, V.: Optimal Control Theory for the Damping of Vibrations of Simple Elastic Systems. Springer, Berlin (1972)
9. Krabs, W.: On Moment Theory and Controllability of One Dimensional Vibrating Systems and Heating Processes. Lecture Notes in Control and Information Sciences. vol. 173. Springer-Verlag, Berlin (1992)
10. Lagnese, J.E.: Boundary Stabilization of Thin Plates. SIAM, Philadelphia (1989)
11. Lagnese, J.E., Leugering, G., Schmidt, E.J.P.G.: Modeling, Analysis and Control of Dynamic Elastic Multi-Link Structures. Springer, New York (1994)
12. Luo, Z.-H., Guo, B.-Z., Morgul, O.: Stability and Stabilization of Infinite Dimensional Systems with Applications. Springer-Verlag, London (1999)
13. Meurer, T.: Control of Higher-Dimensional PDEs. Springer, Berlin (2013)
14. Nabiullin, M.K.: Stationary Motions and Stability of Elastic Satellites (in Russian). Nauka, Novosibirsk (1990)
15. Oostveen, J.: Strongly Stabilizable Distributed Parameter Systems. SIAM, Philadelphia (2000)
16. Sirazetdinov, T.K.: Stability of Systems with Distributed Parameter (in Russian). Nauka, Novosibirsk (1987)

17. Lyapunov, A.M.: The General Problem of the Stability of Motion. (A. T. Fuller trans.) Taylor & Francis, London (1992)
18. Rumyantsev, V.V., Oziraner, A.S.: Stability and Stabilization of Motion with Respect to Part of Variables (in Russian). Nauka, Moscow (1987)
19. Vorotnikov, V.I.: Partial Stability and Control. Birkhäuser, Boston (1998)
20. Cavallo A., De Maria G., Natale C., Pirozzi S.: Active Control of Flexible Structures. From Modeling to Implementation. Springer, London (2010)
21. Rudolph, J.: Flatness Based Control of Distributed Parameter Systems. Shaker Verlag, Aachen (2003)
22. Sira-Ramírez, H., Agrawal, S.K.: Differentially Flat Systems. Taylor & Francis (2004)
23. Tucsnak, M., Weiss, G.: Observation and Control for Operator Semigroups. Birkhäuser, Basel (2009)

Contents

Chapter 1
Basic Results from the Theory of Continuous Semigroups of Operators

Abstract This chapter presents a brief survey of basic results from the theory of C_0-semigroups of operators in a Banach space. Some facts concerning nonlinear semigroups of contractions and their infinitesimal generators are considered as well. The issue of the relative compactness of trajectories is set up for abstract differential equations with accretive operators. We will use these results in order to establish the well-posedness of mathematical models of flexible structures in the next chapters.

To describe the motion of mechanical systems with distributed parameters, it is necessary to develop an appropriate mathematical formalism for the representation of solutions to partial differential equations or infinite systems of ordinary differential equations on an infinite time interval $t \geq 0$. A classical approach to stability problems is presented in the monograph by Daleckii and Kreĭn [1] for solutions to linear and quasi-linear differential equations in a Banach space. It is assumed that the operator in the linear part of the corresponding differential equation is *bounded*. Under this assumption, theorems on stability and instability with respect to the linear approximation is proved in a Banach space.

However, as the dynamics of distributed parameter systems is usually described by partial differential equations, the representation of these processes in infinite-dimensional spaces is associated with the use of *unbounded* differential operators (see, e.g., [2, 3]). This means that results of [1] and other similar studies with bounded generating operators are not directly applicable for the analysis of flexible structures. It is a well-known fact [4, 5], that the issues of existence, uniqueness, and continuous dependence of solutions to the Cauchy problem become crucial for differential equations with unbounded operators. An effective approach in this area is associated with the use of the theory of continuous semigroups of operators in Banach spaces. This approach allows us to prove the well-posedness of the Cauchy problem by exploiting spectral properties of the corresponding infinitesimal operator.

In this chapter, some basic results from the theory of C_0-semigroups of linear operators are presented. These results can be found in classical monographs by Yosida [6], Hille and Phillips [4], Kreĭn [5], Balakrishnan [7], Kato [8], Pazy [9], Fattorini [10].

© Springer International Publishing Switzerland 2015

A.L. Zuyev, *Partial Stabilization and Control of Distributed Parameter Systems with Elastic Elements*, Lecture Notes in Control and Information Sciences 458, DOI 10.1007/978-3-319-11532-0_1

1.1 Strongly Continuous Semigroups of Linear Operators

Let E be a Banach space with the norm $\| \cdot \|$, and let $A : D(A) \to E$ be a linear operator defined on a dense linear subspace $D(A) \subset E$. Consider the following differential equation

$$\frac{\mathrm{d}x(t)}{\mathrm{d}t} = Ax(t) \tag{1.1}$$

with respect to the function $x(t) \in E$ depending on $t \in \mathscr{I}$. The interval $\mathscr{I}$ may be of forms $[0, T]$, $[0, T)$, or $\mathbb{R}^+ = [0, +\infty)$.

Definition 1.1 [5] A function $x : \mathscr{I} \to E$ is a *solution* of Eq. (1.1) on $\mathscr{I}$ if:

(1) $x(t) \in D(A)$ for all $t \in \mathscr{I}$;
(2) the strong derivative $\frac{\mathrm{d}}{\mathrm{d}t}x(t)$ of $x(t)$ exists at each $t \in \mathscr{I}$;
(3) Eq. (1.1) is satisfied for all $t \in \mathscr{I}$.

Solutions in the sense of Definition 1.1 are called *classical solutions* of differential equation (1.1). The *Cauchy problem* on $\mathscr{I}$ consists in defining a solution of Eq. (1.1) on $\mathscr{I}$ that satisfies the initial condition

$$x(0) = x_0 \in D(A). \tag{1.2}$$

Definition 1.2 [5] The Cauchy problem (1.1) and (1.2) is *well-posed* on $\mathscr{I}$ if:

(1) for each $x_0 \in D(A)$, there is a unique solution of the problems (1.1) and (1.2) on $\mathscr{I}$;
(2) if $x_n(t)$ are solutions of (1.1) such that $\|x_n(0)\| \to 0$, $x_n(0) \in D(A)$, $n = 1$, $2, \ldots$, then

$$\lim_{n \to \infty} \|x_n(t)\| = 0, \quad \forall t \in \mathscr{I}.$$

Remark 1.1 If the Cauchy problem (1.1) and (1.2) is well-posed on a segment $[0, T]$, $T > 0$, then it is also well-posed on $\mathbb{R}^+$ [5].

If the Cauchy problem (1.1) and (1.2) is well-posed on $\mathbb{R}^+$ then, for each $t \geq 0$, one can define the operator $S(t)$ that maps $x_0 \in D(A)$ to the value $x(t)$ of the solution of (1.1) and (1.2) at time t:

$$x(t) = S(t)x_0, \quad t \in \mathbb{R}^+. \tag{1.3}$$

This operator $S(t) : D(A) \to D(A)$ is linear due to the linearity of differential equation (1.1) and property (1) from Definition 1.2. Moreover, property (2) of Definition 1.2 implies that $S(t)$ is continuous in $D(A)$. As $D(A)$ is dense in E then $S(t)$ can be extended to a bounded linear operator on E, which we will also denote by $S(t)$.

Thus, for a well-posed Cauchy problem (1.1) and (1.2) on $\mathbb{R}^+$, there exists a one parameter family of bounded linear operators $S(t) : E \to E$, $t \geq 0$, such that formula (1.3) defines the classical solution for $x_0 \in D(A)$. For an arbitrary $x_0 \in E$, we call $x(t) = S(t)x_0$ a *mild solution* of differential equation (1.1) on $\mathbb{R}^+$ [5, 9]. It can be shown that, for any $x_0 \in E$, the function $x(t) = S(t)x_0$ is the limit of a sequence of classical solutions of Eq. (1.1) on $(0, +\infty)$ [5].

If we strengthen the well-posedness condition to the uniform well-posedness then mild solutions on $\mathbb{R}^+$ can be represented in terms of a C_0-semigroup of linear operators. Let us recall the following definitions.

Definition 1.3 [5] A well-posed Cauchy problem (1.1) and (1.2) is said to be *uniformly well-posed* on $\mathbb{R}^+$ if $x_n(0) \to 0$ implies $x_n(t) \to 0$ for solutions of Eq. (1.1), uniformly on each segment $t \in [0, T]$.

Definition 1.4 [9, 11] A one parameter family $\{S(t)\}_{t \geq 0}$ of bounded linear operators $S(t) : E \to E$ is a *strongly continuous semigroup of bounded linear operators* (or a C_0-semigroup) on E if

(i) $S(0) = I$;
(ii) $S(t + s) = S(t)S(s)$ for every $t, s \geq 0$;
(iii) $\lim_{t \to +0} S(t)x = x$ for each $x \in E$.

Here and in the sequel, I denotes the identity operator on E.

Definition 1.5 [5, 9] *The infinitesimal generator* of a semigroup $\{S(t)\}_{t \geq 0}$ is the linear operator $A : D(A) \to E$ defined by

$$Ax = \lim_{t \to +0} \frac{S(t)x - x}{t}, \quad x \in D(A), \tag{1.4}$$

where the domain $D(A)$ of A consists of all $x \in E$ such that the limit (1.4) exists.

Definition 1.6 An operator $A : D(A) \to E$ is *closed* if, for every sequence $\{x_n\} \subset D(A)$ such that $x_n \to x$ and $Ax_n \to y$ as $n \to \infty$, it holds that $x \in D(A)$ and $y = Ax$.

Proposition 1.1 [5] *If the Cauchy problem for (1.1) is uniformly well-posed on $\mathbb{R}^+$ then the corresponding family of solution operators $S(t) : E \to E$ of form (1.3) is a C_0-semigroup on E.*

Proposition 1.2 [5] *Let $A : D(A) \to E$ be a closed linear operator. The Cauchy problem (1.1) and (1.2) is uniformly well-posed on $\mathbb{R}^+$ if and only if A is the infinitesimal generator of a C_0-semigroup of bounded linear operators on E.*

Proposition 1.3 [9] *Let $\{S(t)\}_{t \geq 0}$ be a C_0-semigroup of bounded linear operators on E. Then there exist constants $\omega \geq 0$ and $M \geq 1$ such that*

$$\|S(t)\| \leq Me^{\omega t}, \quad \forall t \geq 0. \tag{1.5}$$

If $\omega = 0$ in the estimate (1.5) then $\{S(t)\}_{t\geq 0}$ is called *uniformly bounded* and if, moreover, $M = 1$ then $\{S(t)\}_{t\geq 0}$ is called *a C_0-semigroup of contractions*. Proposition 1.3 implies the following corollary [9].

Proposition 1.4 *If $\{S(t)\}_{t\geq 0}$ is a C_0-semigroup on E then, for every $x \in E$, $t \mapsto S(t)x$ is a continuous function from $\mathbb{R}^+$ into E.*

According to Proposition 1.2, the well-posedness of the Cauchy problem for differential equation (1.1) is equivalent to the property that A is the infinitesimal generator of a C_0-semigroup. A complete characterization of infinitesimal generators for C_0-semigroups of contractions was proposed by Hille and Yosida in 1948. In order to state the main result in this area, we recall the notion of the resolvent of a linear operator $A : D(A) \to E$ with the domain $D(A) \subset E$.

A number $\lambda \in \mathbb{C}$ is said to be a *regular value* of A, if the equation

$$Ax - \lambda x = y \quad (x \in D(A), y \in E)$$

is solvable with respect to x for each y in a dense subspace of E, and $(A - \lambda I)^{-1}$ is a bounded linear operator. The set of all regular values $\rho(A)$ is called *the resolvent set* of A. *The spectrum* of A is the complement of the resolvent set: $\sigma(A) = \mathbb{C} \setminus \rho(A)$. If A is closed then, for each $\lambda \in \rho(A)$, the bounded linear operator $(A - \lambda I)^{-1}$ is defined on E. This operator $(A - \lambda I)^{-1} : E \to E$ is called *the resolvent* of A.

Theorem 1.1 (The Hille–Yosida Theorem [9]) *A linear (unbounded) operator $A : D(A) \to E$ is the infinitesimal generator of a C_0-semigroup of contractions $\{S(t)\}_{t\geq 0}$ on E if and only if:*

1. *A is closed and $\overline{D(A)} = E$;*
2. *the resolvent set $\rho(A)$ of A contains the interval $(0, +\infty)$, and*

$$\|(A - \lambda I)^{-1}\| \leq \frac{1}{\lambda} \quad \text{for every } \lambda > 0.$$

A modification of this result can be used to describe generators of arbitrary C_0-semigroups.

Theorem 1.2 [5, 9] *A linear (unbounded) operator $A : D(A) \to E$ is the infinitesimal generator of a C_0-semigroup $\{S(t)\}_{t\geq 0}$ satisfying $\|S(t)\| \leq M e^{\omega t}$ on E, if and only if:*

1. *A is closed and $\overline{D(A)} = E$;*
2. *the resolvent set $\rho(A)$ of A contains an interval $(\omega, +\infty)$, and*

$$\|(A - \lambda I)^{-n}\| \leq \frac{M}{(\lambda - \omega)^n} \quad \text{for } \lambda > \omega, \ n = 1, 2, \ldots$$

Verification of the conditions of the Hille–Yosida theorem presents certain technical difficulties in applications of the theory of semigroups to the study of the motion of complex mechanical systems. Another approach to the description of the infinitesimal generator of a contraction semigroup is associated with the dissipativity property.

To state this approach, we consider a particular case of differential equation (1.1) in a Hilbert space H, i.e. $E = H$ in the remaining part of this section. Let us denote by $\langle \cdot, \cdot \rangle$ the inner product in H and introduce the following definitions.

Definition 1.7 A linear operator $A : D(A) \to H$ is said to be *dissipative* if

$$\text{Re}\langle Ax, x \rangle \leq 0, \quad \forall x \in D(A).$$

Definition 1.8 [9, p. 81] A dissipative operator $A : D(A) \to H$ is called *m-dissipative* if

$$\mathscr{R}(I - A) = H.$$

Here, and in the sequel, $\mathscr{R}(I - A)$ stands for the range of $I - A$. If an operator A is dissipative then, obviously, μA is also dissipative for each $\mu > 0$. Thus, for any m-dissipative operator A in H, we have $\mathscr{R}(\lambda I - A) = H$ for each $\lambda > 0$. The following result establishes the relation between dissipative operators and semigroups of contractions.

Theorem 1.3 (Lumer–Phillips Theorem [9, 12]) *Let $A : D(A) \to H$ be a linear operator with dense domain $D(A)$ in H.*

1. *If A is dissipative and there is a $\lambda_0 > 0$ such that*

$$\mathscr{R}(\lambda_0 I - A) = H,$$

   then A is the infinitesimal generator of a C_0-semigroup of contractions $\{S(t)\}_{t \geq 0}$ on H.
2. *If A is the infinitesimal generator of a C_0-semigroup of contractions $\{S(t)\}_{t \geq 0}$ on H then A is dissipative and $\mathscr{R}(\lambda I - A) = H$ for all $\lambda > 0$.*

The above result was proved by Phillips [12] for the case of a Hilbert space, and generalized by Lumer and Phillips [13] for Banach spaces. The Lumer–Phillips theorem can be formulated in terms of m-dissipative operators as follows: a densely defined operator $A : D(A) \to H$ is the infinitesimal generator of a C_0-semigroup if and only if A is m-dissipative.

We will use an important corollary of the Lumer–Phillips theorem by exploiting the dissipativity property of the adjoint operator A^*.

Theorem 1.4 [5, 9] *Let $A : D(A) \to H$ be a densely defined closed linear operator. If both A and A^* are dissipative, then A is the infinitesimal generator of a C_0-semigroup of contractions $\{S(t)\}_{t \geq 0}$ on H.*

If $A : D(A) \to E$ is the infinitesimal generator of a C_0-semigroup on E, then such a semigroup may be denoted as $\{e^{tA}\}_{t \geq 0}$ by the analogy with solutions of linear ordinary differential equations.

If $A : E \to E$ is a bounded linear operator, then $e^{tA} : E \to E$ is defined by the power series:

$$e^{tA} = \sum_{n=0}^{\infty} \frac{1}{n!} (tA)^n.$$

The operator e^{tA} is an analytic function of t, and any solutions of the Cauchy problem (1.1), $x(0) = x_0 \in E$ is represented by the formula $x(t) = e^{tA} x_0$ (cf. [5]).

1.2 Nonlinear Semigroups

Some of the above results of the theory of C_0-semigroups of linear operators can be generalized to represent solutions of nonlinear differential equations. In this section, we present a brief exposition of the theory of nonlinear semigroup of contractions based on the monograph by Barbu [11].

Let X be a closed subset of a real Banach space E. Consider a nonlinear differential equation

$$\frac{dx(t)}{dt} + \Gamma x(t) = 0 \tag{1.6}$$

and initial condition

$$x(0) = x_0 \in X. \tag{1.7}$$

It is assumed that the nonlinear operator $\Gamma : D(\Gamma) \to E$ has dense domain of definition $D(\Gamma)$ in X. As in Definition 1.1, a classical solution of Eq. (1.6) on $\mathbb{R}^+$ is a function $x(t) \in D(\Gamma)$ such that the strong derivative $\frac{d}{dt} x(t)$ exists and satisfies Eq. (1.6) for all $t \geq 0$. In order to characterize properties of the solutions to the Cauchy problem (1.6) and (1.7) for $t \in \mathbb{R}^+$, we generalize Definition 1.4 for nonlinear semigroups (see, e.g., [14, 15]).

Definition 1.9 [14] A *nonlinear semigroup* of operators on X is a one parameter family of mappings $\{S(t) \mid t \in \mathbb{R}^+\}$ from X into X such that

(i) $S(0)x = x$, for each $x \in X$;
(ii) $S(t + s)x = S(t)S(s)x$, for $t, s \geq 0$ and $x \in X$;
(iii) the function $t \mapsto S(t)x$ is continuous in $t \geq 0$ for each $x \in X$.

A nonlinear semigroup $\{S(t)\}$ is *jointly continuous* (or simply *continuous* [15, Chap. 1]) if the mapping $(t, x) \mapsto S(t)x$ from $\mathbb{R}^+ \times X \to X$ is continuous.

Propositions 1.3 and 1.4 imply that, for each C_0-semigroup $\{S(t)\}$ on E, the mapping $(t, x) \mapsto S(t)x$ is continuous with respect to $(t, x) \in \mathbb{R}^+ \times E$. Thus, each C_0-semigroup of bounded linear operators is also a continuous semigroup in the sense of Definition 1.9 with $X = E$. As the estimate (1.5) does not hold for the general nonlinear case, we introduce some important classes of nonlinear semigroups as follows.

Definition 1.10 [11] A nonlinear semigroup $\{S(t)\}_{t \geq 0}$ on X is a *semigroup of contractions* if

$$\|S(t)x_1 - S(t)x_2\| \leq \|x_1 - x_2\|, \quad \forall t \geq 0, \ x_1, x_2 \in X. \tag{1.8}$$

More generally, if instead of (1.8) we have

$$\|S(t)x_1 - S(t)x_2\| \leq e^{\omega t} \|x_1 - x_2\|, \quad \forall t \geq 0, \ x_1, x_2 \in X,$$

then $\{S(t)\}$ is said to be a continuous *ω-quasicontractive semigroup* on X.

It is easy to see that, for any ω-quasicontractive semigroup $\{S(t)\}_{t \geq 0}$ of nonlinear operators on X, the continuity of the mapping $(t, x) \mapsto S(t)x$ follows from its continuity with respect to $t \in \mathbb{R}^+$ for a fixed $x \in X$.

As in Definition 1.5, the *infinitesimal generator of a nonlinear semigroup* $\{S(t)\}_{t \geq 0}$ is the operator

$$F : D(F) \subset X \to E,$$

such that

$$Fx = \lim_{t \to +0} \frac{S(t)x - x}{t}, \quad x \in D(F), \tag{1.9}$$

where the domain of definition $D(F)$ contains all $x \in X$ for which the limit (1.9) exists.

In order to characterize infinitesimal generators of nonlinear semigroups with more generality, we introduce a multi-valued extension $\tilde{\Gamma}$ of the operator Γ in the right-hand side of Eq. (1.6). Namely, let use denote by 2^E the set of all subsets of E, and let us consider a *multi-valued operator* $\tilde{\Gamma} : D(\tilde{\Gamma}) \to 2^E$ with the domain of definition $D(\tilde{\Gamma}) \subset X$. It is obvious that any nonlinear operator $\Gamma : D(\Gamma) \to E$ may be treated as a particular case of $\tilde{\Gamma}$ with $\tilde{\Gamma}x = \{\Gamma x\} \subset E$ and $x \in D(\Gamma) = D(\tilde{\Gamma})$. If the operators Γ and $\tilde{\Gamma}$ are related in this way, then we rewrite Eq. (1.6) as the differential inclusion

$$\frac{dx(t)}{dt} + \tilde{\Gamma}x(t) \ni 0 \tag{1.10}$$

or $\dot{x}(t) \in -\tilde{\Gamma}x(t)$. In general, if $D(\Gamma) \subset D(\tilde{\Gamma})$ and $\Gamma x \in \tilde{\Gamma}x$ for all $x \in D(\Gamma)$, then each solution $x(t)$ of (1.6) also satisfies (1.10). As we will see, the accretivity property of $\tilde{\Gamma}$ plays a crucial role in the study of the generation of nonlinear semigroups. Let us recall some facts from the functional analysis.

For a Banach space E, we denote by E^* its *dual space*, i.e. E^* consists of all continuous linear functionals $x^* : E \to \mathbb{R}$ and is equipped with the dual norm defined by

$$\|x^*\| = \sup_{\|x\| \leq 1} |x^*(x)|.$$

We will also use the notation (x, x^*) for the value $x^*(x)$ of the functional $x^* \in E^*$ at $x \in E$. As E^* is a Banach space, we may introduce the *bidual space* $E^{**} = (E^*)^*$ which is dual to E^*. Each element $x \in E$ generates a continuous linear functional $f(x) : E^* \to \mathbb{R}$ by the formula

$$f(x)(x^*) = x^*(x).$$

This construction defines the evaluation map $f : E \to E^{**}$ acting as follows

$$E \ni x \mapsto f(x) \in E^{**}.$$

If the above defined evaluation map $f : E \to E^{**}$ is surjective, the space E is called *reflexive*. For reflexive Banach spaces, we will identify the elements $x \in E$ and $f(x) \in E^{**}$, so that $E = E^{**}$. It can be shown that all Hilbert spaces as well as the Banach spaces $L^p(0, 1)$ and ℓ^p are reflexive for $p > 1$ (see, e.g., [5, 6]).

A Banach space E is called *strictly convex* if the unit ball B of E is strictly convex, i.e. the boundary ∂B does not contain any line segment. The space E is called *uniformly convex* if, for any $\varepsilon \in (0, 2)$, there exists a $\delta(\varepsilon) > 0$ such that $\|x\| = \|y\| = 1$ and $\|x - y\| \geq \varepsilon$ imply $\|x + y\| \leq 2(1 - \delta(\varepsilon))$. Each uniformly convex space E is also strictly convex. The Hilbert spaces and all $L^p(\Omega)$ spaces, $1 < p < \infty$, are uniformly convex [11].

For any $x \in E$, we define the duality set $J(x) \subset E^*$ as follows:

$$J(x) = \{x^* \in E^* \mid (x, x^*) = \|x\|^2 = \|x^*\|^2\}. \tag{1.11}$$

Definition 1.11 [11] A multivalued operator $\tilde{\Gamma} : D(\tilde{\Gamma}) \to 2^E$ with the domain of definition $D(\tilde{\Gamma}) \subset E$ is called *accretive* if, for every $x_1, x_2 \in D(\tilde{\Gamma})$, $y_1 \in \tilde{\Gamma}x_1$, and $y_2 \in \tilde{\Gamma}x_2$, there is a $w \in J(x_1 - x_2)$ such that

$$(y_1 - y_2, w) \geq 0. \tag{1.12}$$

The accretivity property (1.12) is equivalent to the following condition [11]:

$$\|x_1 - x_2\| \leq \|x_1 - x_2 + \lambda(y_1 - y_2)\|, \quad \forall \lambda > 0, \, x_i \in D(\tilde{\Gamma}), \, y_i \in \tilde{\Gamma}x_i, \, i = 1, 2. \tag{1.13}$$

Definition 1.12 [11] A multivalued operator $\tilde{\Gamma} : D(\tilde{\Gamma}) \to 2^E$, $D(\tilde{\Gamma}) \subset E$ is said to be *dissipative* if $-\tilde{\Gamma}$ is accretive. A multivalued operator $\tilde{\Gamma} : D(\tilde{\Gamma}) \to 2^E$ is called *ω-accretive*, $\omega \in \mathbb{R}$, if $\tilde{\Gamma} + \omega I$ is accretive.

The graph of a multivalued operator $\tilde{\Gamma} : D(\tilde{\Gamma}) \to 2^E$ is defined as

$$\text{graph}(\tilde{\Gamma}) = \{(x, y) \mid x \in D(\tilde{\Gamma}), \ y \in \tilde{\Gamma}x\}. \tag{1.14}$$

According to the standard terminology [11], the notions of accretivity, ω-accretivity, and dissipativity of a *multivalued operator* $\tilde{\Gamma}$ may be equivalently defined for its graph (1.14) as a *subset* of $E \times E$.

Let E be a reflexive Banach space. For a multivalued accretive operator $\tilde{\Gamma} : D(\tilde{\Gamma}) \to 2^E$, $D(\tilde{\Gamma}) \subset E$, we denote by $\Gamma^0 x$ the element of minimum norm in $\tilde{\Gamma}x$. Thus, we consider the map

$$x \in D(\tilde{\Gamma}) \mapsto \Gamma^0 x \in \tilde{\Gamma}x : \|\Gamma^0 x\| = \min_{y \in \tilde{\Gamma}x} \|y\|.$$

If E is a strictly convex space, then the above operator $\Gamma^0 : D(\tilde{\Gamma}) \to E$ is defined in a unique way and is called the *minimal section* of $\tilde{\Gamma}$ [11]. In particular, if $\tilde{\Gamma}x = \{\Gamma x\}$ is a singleton for each $x \in D(\tilde{\Gamma}) = D(\Gamma)$, then $\Gamma^0 = \Gamma$.

The following result is a partial extension of the Lumer–Phillips theorem (Theorem 1.3) to nonlinear semigroups of contractions.

Theorem 1.5 [11] *Let E and E^* be uniformly convex Banach spaces, and let $\tilde{\Gamma} : D(\tilde{\Gamma}) \to 2^E$ be an ω-accretive multivalued operator such that its graph is closed in $E \times E$, and*

$$\overline{\text{conv}\, D(\tilde{\Gamma})} \subset \bigcup_{0 < \lambda < \lambda_0} \mathscr{R}(I + \lambda \tilde{\Gamma}) \ \ \text{for some } \lambda_0 > 0.$$

Then there is a continuous ω-quasicontractive semigroup $\{S(t)\}_{t \geq 0}$ on $\overline{D(\tilde{\Gamma})}$ whose infinitesimal generator F coincides with $-\Gamma^0$.

Here conv $D(\tilde{\Gamma})$ denotes the convex hull of $D(\tilde{\Gamma})$, Γ^0 is the minimal section of $\tilde{\Gamma}$, and the bar denotes the closure operation in E.

1.3 Asymptotic Behavior of Contraction Semigroups

This section presents techniques for the compactness analysis of the trajectories based on the paper by Dafermos and Slemrod [16]. Some related results for monotone operators in a Hilbert space are available in the paper by Bresis [17].

Let us assume that $-\Gamma : D(\Gamma) \to E$ is the infinitesimal generator of a nonlinear semigroup of contractions $\{S(t)\}_{t \geq 0}$ on X (in the sense of Definitions 1.9 and 1.10).

Then, for any $x_0 \in X$, the mild solution of the Cauchy problem (1.6) and (1.7) is represented by the formula

$$x(t) = S(t)x_0, \quad t \in [0, +\infty).$$

In order to study the asymptotic behavior of solutions $x(t)$ as $t \to +\infty$, we first consider the issue of compactness of the trajectory

$$\gamma(x_0) = \{S(t)x_0 \mid t \geq 0\}.$$

Let us recall that the set $\gamma(x_0) \subset X$ is said to be *relatively compact* (or *precompact*), if its closure $\overline{\gamma(x_0)}$ is compact in X [6]. As it will be shown in the next Chapter, the limit behavior of $x(t)$ may be characterized for semigroups with precompact trajectories by using an extension of the Barbashin–Krasovskii theorem [18] or LaSalle's invariance principle [19]. Sufficient conditions for the relative compactness of the trajectories $\gamma(x_0)$ have been proposed in the paper by Dafermos and Slemrod [16]. We formulate here the precompactness result for differential inclusion (1.10) in a Banach space E.

Theorem 1.6 (Dafermos–Slemrod Theorem [16]) *Let $\tilde{\Gamma} : D(\tilde{\Gamma}) \to 2^E$, $D(\tilde{\Gamma}) \subset E$ be an accretive operator such that the operator $(I + \lambda\tilde{\Gamma})^{-1}$ is compact for some $\lambda > 0$, $0 \in \mathscr{R}(\tilde{\Gamma})$, and*

$$\overline{D(\tilde{\Gamma})} \subset \mathscr{R}(I + \lambda\tilde{\Gamma})$$

if $\lambda > 0$ is small enough.
   Assume, moreover, that Γ^0 is the minimal section of $\tilde{\Gamma}$ and that $-\Gamma^0$ is the infinitesimal generator of a nonlinear semigroup of contractions $\{S(t)\}_{t \geq 0}$ on $X = \overline{D(\tilde{\Gamma})}$. Then each trajectory $\gamma(x) = \{S(t)x \mid t \geq 0\}$, $x \in X$ is precompact in E.

References

1. Daletskii, I.L., Kreĭn, M.G.: Stability of Solutions of Differential Equations in Banach Space. AMS, New York (1974)
2. Mikhajlov, V.P.: Partial differential equations. Translated from the Russian by P.C. Sinha. Revised from the 1976 Russian ed. Mir Publishers, Moscow (1978)
3. Oostveen, J.: Strongly Stabilizable Distributed Parameter Systems. SIAM, Philadelphia (2000)
4. Hille, E., Phillips, R.S.: Functional Analysis and Semi-groups. AMS, New York (1957)
5. Kreĭn, S.G.: Linear Differential Equations in Banach Space. AMS, New York (1971)
6. Yosida, K.: Functional Analysis, 6th edn. Springer, Berlin (1995)
7. Balakrishnan, A.V.: Applied Functional Analysis, 2nd edn. Springer, New York (1981)
8. Kato, T.: Perturbation Theory for Linear Operator, 2 Reprint edn. Springer, New York (2013)
9. Pazy, A.: Semigroups of Linear Operators and Applications to Partial Differential Equations. Springer-Verlag, New York (1983)
10. Fattorini, H.O.: Infinite Dimensional Optimization and Control Theory. Cambridge University Press, Cambridge (1999)

11. Barbu, V.: Analysis and Control of Nonlinear Infinite Dimensional Systems. Academic Press, San Diego (1992)
12. Phillips, R.S.: Dissipative operators and hyperbolic systems of partial differential equations. Trans. Amer. Math. Soc. **90**, 193–254 (1959)
13. Lumer, G., Phillips, R.S.: Dissipative operators in a Banach space. Pacific J. Math. **11**, 679–698 (1961)
14. Lakshmikantham, V., Leela, S.: Nonlinear Differential Equations in Abstract Spaces. Pergamon Press, Oxford (1981)
15. Ladyzhenskaya, O.: Attractors for Semigroups and Evolution Equations. Cambridge University Press, Cambridge (1991)
16. Dafermos, C.M., Slemrod, M.: Asymptotic behavior of nonlinear contraction semigroups. J. Funct. Anal. **13**, 97–106 (1973)
17. Bresis, H.: Monotonicity methods in Hilbert spaces and some applications to nonlinear partial differential equations. In: Zarantonello, E.H. (ed.) Contributions to Nonliear Functional Analysis, pp. 101–156. Academic Press, New York (1971)
18. Krasovskii, N.N.: Problems of the Theory of Stability of Motion. Stanford University Press, California (1963)
19. LaSalle, J.P.: Dynamical systems. In: Cesari, L., Hale, J.K., LaSalle, J.P. (eds.) Stability Theory and Invariance Principles, pp. 211–222. Academic Press, New York (1976)

Chapter 2
Partial Asymptotic Stability

Abstract A class of abstract dynamical systems with multivalued flows of solutions in a metric space is introduced in this chapter. For this class of systems, the property of partial asymptotic stability with respect to a continuous functional is studied. In order to characterize the limit set of a trajectory of a multivalued system, a modification of the invariance principle is proposed. This result is applied to derive sufficient conditions for partial asymptotic stability of an equilibrium by using a continuous Lyapunov functional. Such conditions are also formulated for particular classes of systems governed by differential inclusions, ordinary differential equations, and nonlinear semigroups in a Banach space. For further applications of these results to the partial stability analysis of nonlinear abstract differential equations, conditions for the relative compactness of trajectories are derived by considering nonlinear perturbations of dissipative operators. The partial stabilization problem is studied by using differentiable Lyapunov functions for control affine systems in a finite-dimensional space. This treatment is illustrated by examples of the attitude stabilization of a satellite controlled by thrust jets or flywheels.

2.1 Partial Stability of Multivalued Dynamical Systems

Let X be a metric space endowed with the distance $\rho : X \times X \rightarrow \mathbb{R}^+$, $\mathbb{R}^+ = [0, +\infty)$. The evolution of abstract dynamical processes on X will be described by functions $x(t) \in X$ defined for $t \in \mathbb{R}^+$. We denote by κ the set of all functions

$$x : \mathbb{R}^+ \rightarrow X,$$

the set of all subsets of κ is denoted by 2^κ. To introduce the notion of a multivalued dynamical system, we associate a set $\pi(x^0) \subset \kappa$ with each $x^0 \in X$.

Definition 2.1 [1] A map $\pi : X \rightarrow 2^\kappa$ is called a *multivalued D-system on X* if:

(A_1) $\pi(x^0) \neq \emptyset$ for all $x^0 \in X$;
(A_2) $x(0) = x^0$ for each $x(\cdot) \in \pi(x^0)$;

© Springer International Publishing Switzerland 2015

A.L. Zuyev, *Partial Stabilization and Control of Distributed Parameter Systems with Elastic Elements*, Lecture Notes in Control and Information Sciences 458, DOI 10.1007/978-3-319-11532-0_2

(A_3) for any $x^0 \in X$, $s \in \mathbb{R}^+$, $x(\cdot) \in \pi(x^0)$, and $z(\cdot) \in \pi(x(s))$, the following conditions are satisfied: $u(\cdot) \in \pi(x(s))$ and $v(\cdot) \in \pi(x^0)$, where $u(t) = x(t + s)$ and

$$v(t) = \begin{cases} x(t), & t \le s, \\ z(t - s), & t > s; \end{cases}$$

(A_4) given $x^0 \in X$, $\varepsilon > 0$, and $T > 0$, there is a $\delta(x^0, \varepsilon, T) > 0$ such that

$$\rho(\tilde{x}^0, x^0) < \delta, \ \tilde{x}(\cdot) \in \pi(\tilde{x}^0) \Rightarrow \inf_{x(\cdot) \in \pi(x^0)} \left(\sup_{t \in [0,T]} \rho(\tilde{x}(t), x(t)) \right) < \varepsilon;$$

(A_5) for any $x^0 \in X$, $T > 0$, and a sequence $\{x_n(\cdot)\}_{n=1}^{\infty} \subset \pi(x^0)$, there exists an $x(\cdot) \in \pi(x^0)$ such that

$$\liminf_{n \to \infty} \left(\sup_{t \in [0,T]} \rho(x_n(t), x(t)) \right) = 0.$$

We will refer to an element $x(t)$ of $\pi(x^0)$ as to *a solution of the initial value problem* $x(0) = x^0$ for π. When considering an autonomous system of differential equations, the above $\pi(x^0)$ represents the set of all solutions for the Cauchy problem on $\mathbb{R}^+$. Assumptions A_1 and A_2 state the global existence property, while A_3 means that the translation of a solution is a solution. Conditions in A_4 and A_5 provide extra regularity properties without assuming the uniqueness of solutions.

Definition 2.2 Let $x(\cdot) \in \pi(x^0)$. An element $q \in X$ is said to be a (*positive*) *limit point of* x if there is a sequence $t_n \to +\infty$ such that $x(t_n) \to q$ as $n \to \infty$. The set of all such limit points is denoted by $\Omega(x)$ and called the (*positive*) *limit set of* x.

Definition 2.3 A set $F \subset X$ is said to be *semi-invariant for* π if, for every $x^0 \in F$, there exists at least one $x(\cdot) \in \pi(x^0)$ such that $x(t) \in F$ for all $t \in \mathbb{R}^+$.

Definition 2.4 We say that $x(\cdot) \in \pi(x^0)$ is *precompact* if $\bigcup_{t \ge 0}\{x(t)\}$ is contained in a (sequentially) compact subset of X.

An important property of the limit sets of the autonomous differential equations is that they are invariant (cf. [2, App. III]). The following lemma extends this well-known result for the class of multivalued D-systems on a metric space.

Lemma 2.1 *Let π be a multivalued D-system and let $x(\cdot) \in \pi(x^0)$. If the trajectory x is precompact then $\Omega(x)$ is nonempty and semi-invariant.*

Proof The precompactness of $\{x(t) \,|\, t \ge 0\}$ implies that, for any sequence $t_n \to +\infty$, the sequence $\{x(t_n)\}_{n=1}^{\infty}$ has a limit point, therefore, $\Omega(x) \ne \emptyset$.

Let us show that, for each $T > 0$ and $x_0^* \in \Omega(x)$, there is a function $\xi(\cdot) \in \pi(x_0^*)$ satisfying the condition $\xi(t) \in \Omega(x)$ for all $t \in [0, T]$. Since $x_0^* \in \Omega(x)$, there exists a sequence $t_n \to +\infty$ such that $x(t_n) \to x_0^*$ as $n \to \infty$. Define the sequence

$\{\phi_n(\cdot)\}_{n=1}^{\infty} \subset \kappa\colon \phi_n(t) = x(t_n + t),\ t \in \mathbb{R}^+$. Then $\phi_n(\cdot) \in \pi(x(t_n))$ because of A_3. Let $\{\phi_{n(k)}(\cdot)\}_{k=1}^{\infty}$ be a subsequence of $\{\phi_n(\cdot)\}_{n=1}^{\infty}$ satisfying the condition $\rho(\phi_{n(k)}(0), x_0^*) < \delta_k$, where the numbers $\delta_k = \delta(x_0^*, 1/k, T) > 0$ are chosen as in A_4. By assumption A_4, there is a sequence $\{\psi_k(\cdot)\}_{k=1}^{\infty} \subset \pi(x_0^*)$ such that

$$\sup_{t \in [0,T]} \rho(\phi_{n(k)}(t), \psi_k(t)) < \frac{1}{k}, \quad k = 1, 2, \dots \qquad (2.1)$$

Then A_5 implies that there exist $\xi(\cdot) \in \pi(x_0^*)$ and a subsequence $\{\psi_{k(m)}(\cdot)\}_{m=1}^{\infty}$:

$$\lim_{m \to \infty} \left(\sup_{t \in [0,T]} \rho(\psi_{k(m)}, \xi(t)) \right) = 0.$$

The above formula together with (2.1) imply

$$\lim_{m \to \infty} \left(\sup_{t \in [0,T]} \rho(\phi_{n(k(m))}, \xi(t)) \right) = 0.$$

Since each $\phi_n(\cdot)$ is the translation of $x(\cdot)$, each value of $\xi(t)$ ($0 \le t \le T$) belongs to $\Omega(x)$.

To conclude the proof, we apply the above construction infinitely many times at the points $x_i^* = \xi_{i-1}(T)$, where $\xi_0(\cdot) = \xi(\cdot)$. As a result, we get the system of functions $\xi_i(\cdot) \in \pi(\xi_{i-1}(T))$ such that $\xi_i(t) \in \Omega(x)$ for all $t \in [0, T]$, $i = 1, 2, \dots$ Then A_3 implies that the function $x^*(t) = \xi_{[t/T]}(\{t/T\}T)$ is an element of $\pi(x_0^*)$, where $[t/T]$ and $\{t/T\}$ denote the integer and the fractional parts of t/T, respectively. Moreover, $x^*(t) \in \Omega(x)$ for all $t \in \mathbb{R}^+$. $\square$

The limit sets of a dynamical system can be characterized in the terms of a Lyapunov function. A powerful machinery in this area is given by the invariance principle that is valid for the abstract systems on a Fréchet space [3]. We prove here a similar proposition for the D-systems in the sense of Definition 2.1.

Lemma 2.2 *Let π be a multivalued D-system, $x(\cdot) \in \pi(x^0)$. Suppose that there exists a continuous map $V : X \to \mathbb{R}^+$ such that $\xi_0 \in X$ and $\xi(\cdot) \in \pi(\xi_0)$ imply $V(\xi(t))$ is non-increasing on $\mathbb{R}^+$. If $x(\cdot)$ is precompact, then*

$$\Omega(x) \subset \{p \in X \mid V(\xi(t)) = c \text{ for some } \xi(\cdot) \in \pi(p),\ t \in \mathbb{R}^+\} \qquad (2.2)$$

for some constant c.

Proof Precompactness of $\bigcup_{t \ge 0}\{x(t)\}$ and continuity of V imply that $\Omega(x) \ne \emptyset$ and $V(x(t))$ is bounded on $\mathbb{R}^+$. Since $V(x(t))$ is non-increasing, there exists the limit

$$\lim_{t \to +\infty} V(x(t)) = c \ne -\infty.$$

As V is continuous, $V(x^*) = c$ for any $x^* \in \Omega(x)$. It means that $\Omega(x)$ is a subset of $\{p \in X \mid V(p) = c\}$. But as $\Omega(x)$ is semi-invariant (Lemma 2.1), for any $p \in \Omega(x)$ there should be a $\xi(\cdot) \in \pi(p)$ such that $\xi(t) \in \Omega(x)$. It implies $V(\xi(t)) = c$ for all $t \in \mathbb{R}^+$. $\square$

Our goal is to apply the invariance principle for the analysis of partial asymptotic stability in abstract spaces. To introduce this notion, let us call $x^0 \in X$ an *equilibrium of π* if the function $x(t) \equiv x^0$ belongs to $\pi(x^0)$.

Definition 2.5 Let π be a multivalued D-system on X, and let $\mu : X \to \mathbb{R}^+$. The equilibrium x^0 of π is said to be *asymptotically stable* with respect to μ if

(i) Given $\varepsilon > 0$ there exists $\delta(\varepsilon) > 0$ such that $\rho(\tilde{x}^0, x^0) < \delta$ implies $\mu(\tilde{x}(t)) < \varepsilon$ for all $\tilde{x}(\cdot) \in \pi(\tilde{x}^0)$ and all $t \in \mathbb{R}^+$.
(ii) There exists a $\Delta > 0$ such that $\rho(\tilde{x}^0, x^0) < \Delta$ implies

$$\lim_{t \to \infty} \mu(\tilde{x}(t)) = 0. \tag{2.3}$$

Remark 2.1 The notion of stability with respect to two metrics was introduced by Movčan in the paper [4]. Our approach differs from Movčan's work as we consider the partial stability for multivalued processes here.

To formulate stability results, we introduce the standard class $\mathcal{K}$ of comparison functions that consists of all continuous strictly increasing functions $\alpha : \mathbb{R}^+ \to \mathbb{R}^+$ such that $\alpha(0) = 0$. Then we prove the following theorem on sufficient conditions of partial asymptotic stability in terms of a continuous Lyapunov functional V on X.

Theorem 2.1 *Let π be a multivalued D-system on a metric space X, and let x^0 be its equilibrium. Assume that there is a pair of continuous functionals μ, $V : X \to \mathbb{R}^+$ satisfying the following conditions.*

C_1. *There exist $\alpha_1(\cdot), \alpha_2(\cdot) \in \mathcal{K}$ such that*

$$\alpha_1(\mu(x)) \le V(x) \le \alpha_2(\rho(x^0, x)) \quad \text{for all } x \in X. \tag{2.4}$$

C_2. *For any $\tilde{x}^0 \in X$, $\tilde{x}(\cdot) \in \pi(\tilde{x}^0)$, the function $V(\tilde{x}(t))$ is non-increasing on $\mathbb{R}^+$.*
C_3. *There exists a $\Delta > 0$ such that $\rho(\tilde{x}^0, x^0) < \Delta$ and $\tilde{x}(\cdot) \in \pi(\tilde{x}^0)$ imply precompactness of $\tilde{x}(\cdot)$.*
C_4. *The set*

$$M_1 = \{p \in X \mid V(\tilde{x}(t)) \text{ is constant on } \mathbb{R}^+$$

$$\text{for some } \tilde{x}(\cdot) \in \pi(p)\} \tag{2.5}$$

is contained in

$$\text{Ker } \mu = \{p \in X \mid \mu(p) = 0\}.$$

Then the equilibrium x^0 is asymptotically stable with respect to μ.

Proof First we prove the property (i) from Definition 2.5 by generalizing Rumyant-sev's theorem [5, Theorem 5.1], [6] on partial stability with respect to a part of the variables. Then we apply Lemma 2.2 to show the property (ii).

Condition C_2 implies $V(\tilde{x}(t)) \le V(\tilde{x}^0)$ for all $\tilde{x}^0 \in X, \tilde{x}(\cdot) \in \pi(\tilde{x}^0)$, and $t \in \mathbb{R}^+$. By combining this inequality with (2.4), we get

$$\mu(\tilde{x}(t)) \le \alpha_1^{-1}\left(\alpha_2\left(\rho\left(\tilde{x}^0, x^0\right)\right)\right), \qquad (2.6)$$

where the function $\alpha_1^{-1}(\tau)$ exists and increases at least for small enough $\tau > 0$, since $\alpha(\cdot) \in \mathcal{K}$. Therefore, the function

$$\gamma(\delta) = \alpha_1^{-1}\left(\alpha_2(\delta)\right)$$

is continuous, nonnegative, and strictly increasing on some interval $[0, \delta^*), 0 < \delta^* \le +\infty$. It means that for arbitrary $\varepsilon > 0$ there exists $\delta \in (0, \delta^*)$ such that $\gamma(\delta) \le \varepsilon$. Hence, if $\rho(\tilde{x}^0, x^0) < \varepsilon$ then (2.6) implies

$$\mu(\tilde{x}(t)) \le \gamma(\rho(\tilde{x}^0, x^0))$$

for all $\tilde{x}(\cdot) \in \pi(\tilde{x}^0)$ and all $t \in \mathbb{R}^+$.

To conclude the proof, it suffices to establish the limit existence for (2.3). Let Δ be chosen as in C_3, and let $\rho(\tilde{x}^0, x^0) < \Delta$. Therefore, for any $\tilde{x}(\cdot) \in \tilde{x}^0$, the set $\Omega(\tilde{x}) \ne \emptyset$ is included in (2.5) because of Lemma 2.2. Condition C_4 implies

$$\Omega(\tilde{x}) \subset \text{Ker } \mu. \qquad (2.7)$$

To show (2.3), let us assume the contrary: there are some $\varepsilon > 0$ and $t_n \to +\infty$ as $n \to \infty$ such that

$$\mu(\tilde{x}(t_n)) > \varepsilon, \quad n = 1, 2, \ldots \qquad (2.8)$$

Since $\tilde{x}(\cdot)$ is precompact, there exists a subsequence $\{t_{n(k)}\}_{k=1}^\infty$ such that $\tilde{x}(t_{n(k)}) \to x^* \in \Omega(\tilde{x})$ as $k \to \infty$. By (2.7), $\mu(x^*) = 0$. Continuity of μ implies

$$|\mu(\tilde{x}(t_{n(k)})) - \mu(x^*)| = \mu(\tilde{x}(t_{n(k)})) \to 0, \quad \text{as } k \to \infty.$$

But the above contradicts to (2.8). Therefore,

$$\lim_{t \to +\infty} \mu(\tilde{x}(t)) = 0$$

for all $\tilde{x}(\cdot) \in \pi(\tilde{x}^0)$ provided that $\rho(\tilde{x}^0, x^0) < \Delta$. $\qquad\square$

2.2 Application to Differential Inclusions and Ordinary Differential Equations

An important application of the concept of partial stability for systems with multi-valued flow comes from differential inclusions and Filippov's approach to ordinary differential equations with discontinuous right-hand sides.

2.2.1 Differential Equations with Discontinuous Right-Hand Sides

Consider a system of ordinary differential equations

$$\dot{x}(t) = f(x(t)), \quad x(t) \in X \subseteq \mathbb{R}^n, \tag{2.9}$$

where the function $f : X \to \mathbb{R}^n$ is assumed to be bounded on each compact subset D of domain X. In the sequel, we assume that $0 \in X$ and $f(0) = 0$, so that system (2.9) admits the trivial solution $x(t) \equiv 0$.

It is a well-known fact [7] that classical solutions of system (2.9) may not exist if f is a discontinuous function. To ensure the existence and extendability of solutions to system (2.9), one should use a generalized notion of solutions which is applicable for differential equations with discontinuous right-hand sides. We use here the following definition of solutions due to Filippov.

Definition 2.6 *A solution of system* (2.9) is an absolutely continuous function $x(t) \in X$, defined for $-\infty \leq T^- < t < T^+ \leq +\infty$, that satisfies the following differential inclusion

$$\dot{x}(t) \in \text{co}\, H(x(t)), \tag{2.10}$$

almost everywhere on $t \in (T^-, T^+)$. For each $x \in X$, the set $H(x) \subset \mathbb{R}^n$ contains $f(x)$ and the set of all limit points of $f(y)$ as $y \to x$. Here $\text{co}\, H(x)$ is the convex hull of $H(x)$.

Definition 2.6 corresponds to the simplest convex definition according to [7, Sect. 4] (see also Proposition 1 in [8, Chap. 2]). It is obvious that each classical solution of (2.9) is also a solution in the sense of Filippov. Definition 2.6 is equivalent to the classical definition of solutions if the function $f : X \to \mathbb{R}^n$ is continuous.

For a given $x^0 \in X$, a solution $x(t)$ to the Cauchy problem for differential inclusion (2.10) with initial data

$$x(0) = x^0 \in X \tag{2.11}$$

may not be unique. For further analysis, we will make an extra assumption.

Assumption 2.1 If $x(t)$ is a solution of differential inclusion (2.10) on $t \in I$ then either $I \supset [0, +\infty)$ or $x(t)$ may be extended to some interval $t \in \tilde{I} \supset [0, +\infty)$.

Under this assumption, we introduce the following set-valued map on X:

$$x^0 \mapsto \pi(x^0) = \{x(\cdot) \mid x(t) \text{ is a solution to the Cauchy problem (2.10), (2.11) on } t \geq 0\}. \tag{2.12}$$

By exploiting regularity results from [7], we show that the above defined map π is a multivalued D-system.

Lemma 2.3 *Let the Assumption 2.1 be satisfied. Then the set-valued map $x^0 \mapsto \pi(x^0)$, given by (2.12), is a multivalued D-system on X in the sense of Definition 2.1.*

Proof As the function $f(x)$ is bounded on each compact subset of X, the set-valued function $F(x) = \operatorname{co} H(x)$, introduced in Definition 2.6, is upper semicontinuous by Lemma 1 of [7, Sect. 6]. Hence, $F(x)$ satisfies the *basic conditions* of [7, Sect. 7], i.e. the set $F(x)$ is nonempty, bounded, and closed for all $x \in X$, and $F(x)$ is upper semicontinuous in x. Then, for each $x^0 \in X$, there is a solution $x(t)$, $t \in [0, T^+)$ of the Cauchy problem (2.10), (2.11) according to Theorem 1 of [7, Sect. 7], and $T^+ = +\infty$ by Assumption 2.1. This implies that the set-valued map $\pi(x^0)$, defined by (2.12), satisfies conditions (A_1) and (A_2) of Definition 2.1. Condition (A_3) also holds as $f(x)$ and $F(x)$ do not depend on t. Condition (A_4) is a consequence of Theorem 1 from [7, Sect. 8] on the dependence of solutions on the initial data. Condition (A_5) follows from the fact that the limit of a uniformly convergent sequence of solutions $\{x_n(t)\}$ of differential inclusion (2.10) is a solution of (2.10) (see Corollary 1 of [7, Sect. 7]). Thus, all the conditions (A_1)–(A_5) of Definition 2.1 are satisfied, and the set-valued map $\pi(x^0)$, introduced in (2.12), is a multivalued D-system on X in the sense of Definition 2.1.

To present a finite-dimensional version of Theorem 2.1, we write the state vector x of system (2.9) as

$$x = (y_1, ..., y_{n_1}, z_1, ..., z_{n_2}), \quad y = (y_1, ..., y_{n_1}) \in \mathbb{R}^{n_1}, \quad z = (z_1, ..., z_{n_2}) \in \mathbb{R}^{n_2},$$
$$n_1 + n_2 = n. \tag{2.13}$$

We also assume that X is a domain of form

$$X = \{x \in \mathbb{R}^n \mid z \in \mathbb{R}^{n_2}, \ \|y\| < N\} \tag{2.14}$$

where $\|\cdot\|$ is the standard Euclidean norm of a vector and N is a positive constant. For a Filippov solution $x(t)$ of system (2.9), we will refer to its y- and z-components as $y(t)$ and $z(t)$, respectively.

Let us recall the following definition of asymptotic stability with respect to a part of variables in the sense of Lyapunov [9] and Rumyantsev [5].

Definition 2.7 [5, 6] The solution $x = 0$ of system (2.9) is *asymptotically y-stable* if, for any $\varepsilon > 0$, there exists a $\delta > 0$ such that any solution $x(t)$ of (2.9) with $\|x(0)\| < \delta$ is defined on $\mathbb{R}^+$, $\|y(t)\| < \varepsilon$ for all $t \geq 0$, and $\|y(t)\| \to 0$ as $t \to +\infty$.

If we treat solutions of system (2.9) in the sense of A.F. Filippov, then it is easy to see that Definition 2.7 is equivalent to Definition 2.5 with

$$\mu(x) = \|y\|,$$

provided that the set-valued map $x^0 \in X \mapsto \pi(x^0)$ is defined by (2.12).

Remark 2.2 The concepts of weak and strong stability of solutions are usually addressed in the theory of differential inclusions [10]. Definition 2.7 is related to the strong partial stability, so that $\|y(t)\| < \varepsilon$ and $y(t) \to 0$ for *each* solution $x(t)$ whenever $\|x(0)\| < \delta$.

For a differentiable function $V(x)$ in X, its upper time derivative along the trajectories of differential inclusion (2.10) is

$$\dot{V}^*(x) = \sup_{p \in \text{co}\, H(x)} \langle \nabla V(x), p \rangle,$$

where $\nabla V(x)$ is the gradient of $V(x)$ and $\langle \cdot, \cdot \rangle$ is the scalar product in $\mathbb{R}^n$.

As a corollary of Theorem 2.1, we obtain the following result.

Theorem 2.2 *Assume that, for some $\Delta > 0$, each Filippov solution $x(t)$ of system (2.9) is bounded for $t \geq 0$ provided that $\|x(0)\| < \Delta$. Let $V \in C^1(\overline{X})$ be a function such that $V(0) = 0$ and the following conditions hold:*

(1) $V(x) \geq \alpha(\|y\|)$ *for some $\alpha \in \mathcal{K}$;*
(2) $\dot{V}^*(x) \leq 0$ *for all $x \in X$;*
(3) *the set $M = \{x \mid y = 0\}$ is invariant for (2.10) with $t \geq 0$;*
(4) *the set $\{x \mid \dot{V}^*(x) = 0\}\backslash M$ does not contain any weakly invariant subset for (2.10) with $t \geq 0$.*

Then the solution $x = 0$ of system (2.9) is asymptotically y-stable.

Proof Let $x(t)$, $t \in I$ be a Filippov solution of (2.9) with $\|x(0)\| < \Delta$. Without loss of generality we assume that $I \supset [0, +\infty)$, otherwise, as each solution is bounded, $x(t)$ may be extended to some interval $\tilde{I} \supset [0, +\infty)$ by Theorem 2 of [7, Sect. 7]. Thus, the Filippov solutions of system (2.9) correspond to a multivalued *D*-system $x^0 \mapsto \pi(x^0)$ on X by Lemma 2.3. The boundedness of the solutions implies that condition C_3 of Theorem 2.1 holds as each bounded subset of a finite dimensional space is precompact. Let us show that condition C_1 of Theorem 2.1 holds with $\alpha_1 = \alpha \in \mathcal{K}$ given in condition (1) and

$$\alpha_2(\rho) = \sup_{\|x\| \leq \rho,\, x \in X} V(x).$$

The above defined α_2 is a function of class $\mathcal{K}$ because $V(0) = 0$ and $V(x)$ is continuous. If $x(t)$ is a solution of differential inclusion (2.10) then

$$
\begin{aligned}
\frac{d}{dt} V(x(t)) &= \lim_{h \to 0} \frac{V(x(t+h)) - V(t)}{h} \\
&= \lim_{h \to 0} \frac{1}{h} \{ \langle \nabla V(x(t)), x(t+h) - x(t) \rangle + o(\|x(t+h) - x(t)\|) \} \\
&= \lim_{h \to 0} \{ \langle \nabla V(x(t)), p \rangle_{p \in \mathrm{co} H(x(t))} + o(1) \} \le \dot{V}^*(x(t)) \le 0
\end{aligned}
$$

almost everywhere on $t \in [0, +\infty)$. This implies that $V(x(t))$ is a non-increasing function of $t \ge 0$, so that condition C_2 of Theorem 2.1 holds.

To prove condition C_4, let us assume the contrary: let $x(t)$, $t \ge 0$ be a solution of differential inclusion (2.10) such that $V(x(t)) \equiv 0$ and $y(\tau) \ne 0$ for some $\tau \ge 0$. The property $V(x(t)) \equiv 0$ together with condition $\dot{V}^*(x) \le 0$ imply that $x(t) \in M_0$ for all $t \ge 0$, where

$$
M_0 = \{ x \in X \mid \dot{V}^*(x) = 0 \}.
$$

As the set $M = \{x \mid y = 0\}$ is invariant [condition (3)], then $y(t) \ne 0$ for all $t \ge 0$ under our assumptions. This implies that $x(t) \in M_0 \backslash M$ for all $t \ge 0$ which contradicts condition (4). This contradiction shows that the assumption C_4 is satisfied, so that the equilibrium $x = 0$ of system (2.9) is asymptotically stable with respect to $\mu(x) = \|y\|$ by Theorem 2.1.

Remark 2.3 In order to formulate sufficient conditions of partial stability, it is natural to assume that the solutions are z-*extendable* [5, 6], i.e. if $x(t)$ is a solution of (2.10) for $T^- < t < T^+ < +\infty$ and $z(t) \to \infty$ as $t \to T^+$ then $\|y(t)\| \to N$ as $t \to T^+$. In Theorem 2.2, such z-extendability assumption follows from the boundedness of the solutions.

Remark 2.4 If the right-hand side of system (2.9) is of class $C^1(X)$, then Theorem 2.2 is equivalent to the Risito–Rumyantsev theorem [11], [5, Theorems 19.1–19.2]. Further on, if $n_1 = n$ so that $y = x$ and $\mu(x) = \|x\|$, then Theorem 2.2 is reduced to the Barbashin–Krasovskii theorem [12] on asymptotic stability of the equilibrium $x = 0$.

For a given function $V(x)$ such that $\dot{V}(x) \le 0$ along the trajectories of system (2.9), the vector of variables y satisfying the property that the solution $x = 0$ of system (2.9) is y-asymptotically stable (in the sense of Definition 2.7) may be defined by using the method of the paper [13].

Remark 2.5 As Theorem 2.2 follows from Theorem 2.1, a part of its proof is actually based on the invariance principle (Lemma 2.2) with a differentiable Lyapunov function $V(x)$. A modification of the invariance principle with a non-smooth Lyapunov function was used for the stability analysis of differential inclusions in the paper [14].

2.3 Stabilization of Finite-Dimensional Systems with Respect to a Part of Variables

Let us consider a control system:

$$\dot{x} = f_0(x) + \sum_{i=1}^{m} u_i f_i(x), \quad x \in X \subset \mathbb{R}^n, \ u = (u_1, \dots, u_m) \in \mathbb{R}^m, \quad (2.15)$$

where x is the state vector and u is the control. We assume that X is a domain of form (2.14), $f_i \in C(X)$ and $f_0(0) = 0$, so that system (2.15) admits the trivial solution $x(t) \equiv 0$ with $u = 0$.

Let us remark, that if system (2.15) has an integral then it is neither controllable nor stabilizable. In this case, only partial stabilization may be possible (see, e.g., [6]).

The goal of this section is to develop an effective strategy for partial stabilization of system (2.15) based on Theorem 2.2.

2.3.1 Theorem on Partial Stabilization

According to notations (2.13), system (2.15) may be written as

$$\dot{y} = f_{01}(y, z) + \sum_{i=1}^{m} u_i f_{i1}(y, z), \quad \dot{z} = f_{02}(y, z) + \sum_{i=1}^{m} u_i f_{i2}(y, z). \quad (2.16)$$

The functions $f_{ij}(y, z)$ are considered in the domain X. By an admissible feedback for system (2.16) we treat any function $k(x) \in C(X)$ such that $k(0) = 0$. We use the following definition.

Definition 2.8 Control system (2.16) is said to be *y-stabilizable* if there exists an admissible feedback law $u = k(x)$ such that the trivial solution of the corresponding closed-loop system is asymptotically *y*-stable.

The consideration of continuous feedback laws simplifies the stability analysis with Theorem 2.2 as the sets of classical and Filippov's solutions of the closed-loop system coincide. However, the multivalued framework of Sect. 2.2 is important for our study as systems with merely continuous right-hand sides may not exhibit the uniqueness of solutions.

Let $V(x)$ be a function of class $C^1(X)$. The time derivative of $V(x)$ along the trajectories of system (2.16) is:

$$\dot{V} = a(x) + \langle u, b(x) \rangle,$$

where

$$a(x) = \langle \nabla V(x), f_0(x) \rangle, \quad b_i(x) = \langle \nabla V(x), f_i(x) \rangle, \quad i = 1, 2, \ldots, m,$$
$$b(x) = (b_1(x), b_2(x), \ldots, b_m(x)).$$
(2.17)

The basic result we shall prove in this section is the following.

Theorem 2.3 *Let $V \in C^1(X)$ be a function such that $V(0) = 0$ and the following conditions hold:*

(1) *$V(x) \geq \alpha(\|y\|)$ for some $\alpha \in \mathcal{K}$;*
(2) *the equation $a(x) + \langle u^0(x), b(x) \rangle = 0$ has a solution $u^0 \in C(X)$ for which the set*

$$M_1 = \{x \mid b(x) = 0, \ y \neq 0\}$$

does not contain any trajectory $\{x(t) \mid t \geq 0\}$ of system (2.16) with the control $u = u^0(x)$;
(3) *there exist a positive number $\Delta > 0$ and a function $h \in C(X)$, $h(x) > 0$ such that each solution $x(t)$ of system (2.16) with the initial condition $\|x(0)\| < \Delta$ and the feedback control*

$$u = u^0(x) - h(x)b(x)$$
(2.18)

has bounded coordinates $z_j(t)$, $1 \leq j \leq n_2$, for all $t \geq 0$.

Then the solution $x = 0$ of the closed-loop system (2.16) with (2.18) is asymptotically y-stable (i.e. system (2.16) is y-stabilizable).

Proof By substituting (2.18) into (2.16) and computing $\dot{V}$, we get

$$\dot{V} = -h \|b(x)\|^2 \leq 0.$$

The time-derivative $\dot{V}$ vanishes on the following set:

$$M = \{x \mid b(x) = 0\}.$$

It is easy to see that $u(x) = u^0(x)$ for each $x \in M$ if the feedback is given by formula (2.18). Therefore, condition (2) implies that the set $M \setminus \{x : y = 0\}$ does not contain any positive semitrajectory for the closed-loop system (2.16) with (2.18).

Condition (3) together with the inequality $\dot{V} \leq 0$ guarantees the boundedness of the solutions starting from Δ-neighborhood of the origin.

Thus, all the conditions of Theorem 2.2 hold for the closed-loop system (2.16) with (2.18).

In the sequel, we apply Theorem 2.3 for studying a couple of examples of single-axis stabilization [15].

2.3.2 Partial Stabilization of a Rigid Body

Consider a model that describes the rotation of a satellite around its center of mass under the action of attitude control thrust jets (Fig. 2.1). We treat the satellite as a rigid body rotating around its center of mass (fixed point O). Let $Oe_1e_2e_3$ be a basis associated with the rigid body, and let v be a unit vector which is fixed in the inertial frame. The equations of motion we can be written in the Euler–Poisson form as follows [16]:

$$\dot{\omega}_1 = \frac{A_2 - A_3}{A_1}\omega_2\omega_3 + u_1, \quad \dot{\omega}_2 = \frac{A_3 - A_1}{A_2}\omega_1\omega_3 + u_2, \quad \dot{\omega}_3 = \frac{A_1 - A_2}{A_3}\omega_1\omega_2,$$
(2.19)

$$\dot{v}_1 = \omega_3 v_2 - \omega_2 v_3, \quad \dot{v}_2 = \omega_1 v_3 - \omega_3 v_1, \quad \dot{v}_3 = \omega_2 v_1 - \omega_1 v_2.$$
(2.20)

Here $\omega = \omega_1 e_1 + \omega_2 e_2 + \omega_3 e_3$ is the angular velocity vector of the rigid body, $v = v_1 e_1 + v_2 e_2 + v_3 e_3$, and A_i is the moment inertia of the rigid body with respect to the axis defined by e_i, $i = 1, 2, 3$. We also assume that the directions of e_i are principal axes of inertia of the body. The action of jet torques is described by control parameters u_1 and u_2.

System (2.19) and (2.20) admits the following particular solution with $u_1 = 0$ and $u_2 = 0$:

$$\omega_1 = \omega_2 = \omega_3 = 0, \quad v_1 = v_2 = 0, \quad v_3 = 1.$$
(2.21)

Fig. 2.1 Satellite with attitude control thrust jets

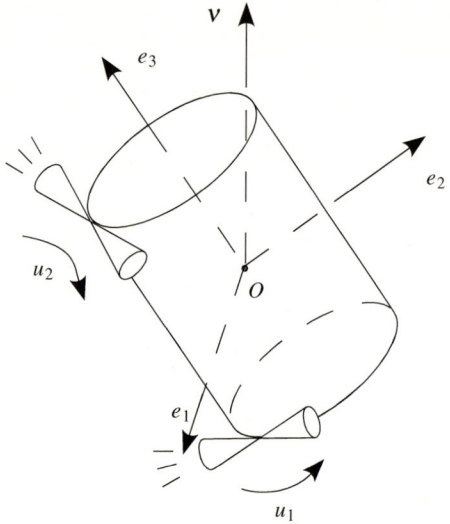

This solution corresponds to the equilibrium for which vectors e_3 and v coincide. Let us remark that solution (2.21) of system (2.19) and (2.20) cannot be made asymptotically stable (with respect all state variables) due to the geometric integral:

$$v_1^2 + v_2^2 + v_3^2 = \text{const.} \tag{2.22}$$

Stabilizability of the Euler equations of form (2.19) has been studied by many authors (see, e.g., [16–18] and references therein). Our investigation is based on the application of Theorem 2.3 in order to stabilize solution (2.21) with respect to the following variables:

$$(\omega_1, \omega_2, v_1, v_2). \tag{2.23}$$

This choice of variables correspond to the stabilization of the third principal axis of inertia (e_3) around the fixed direction v. So, v_1 and v_2 and their derivatives are required to be "small" and tending to zero as $t \to +\infty$, while the other ones are required to be merely bounded.

We define a Lyapunov function candidate as follows:

$$2V = A_1\omega_1^2 + A_2\omega_2^2 + \varkappa(v_1^2 + v_2^2), \quad \varkappa > 0.$$

A straightforward application of formulas (2.17) yields

$$a(x) = (A_2 - A_1)\omega_1\omega_2\omega_3 + \varkappa v_3(\omega_1 v_2 - \omega_2 v_1),$$
$$b_1(x) = A_1\omega_1, \quad b_2(x) = A_2\omega_2. \tag{2.24}$$

A particular solution of the equation $a(x) + \langle u^0, b(x) \rangle = 0$ can be taken in the form

$$u_1^0(x) = \omega_2\omega_3 - \frac{\varkappa}{A_1} v_2 v_3, \quad u_2^0(x) = -\omega_1\omega_3 + \frac{\varkappa}{A_2} v_1 v_3. \tag{2.25}$$

It is easy to check that all trajectories of system (2.19) and (2.20) with control (2.25) satisfy the condition $v_1 = v_2 = 0$ on the set

$$M_1 : A_1\omega_1 = A_2\omega_2 = 0,$$

provided that the initial value is taken in some neighborhood of solution (2.21). This proves that condition (2) of Theorem 2.3 holds.

All solutions of system (2.19) and (2.20) are bounded with respect to v_i because of integral (2.22). So, it suffices to ensure the boundedness of the solutions with respect to ω_3. In order to prove the boundedness, we apply Theorem 39.1 from the monograph [5] with the following function:

$$2W = A_1\omega_1^2 + A_2\omega_2^2 + A_3\omega_3^2 + \varkappa(v_1^2 + v_2^2).$$

The time-derivative of $W(x)$ along the trajectories of the closed-loop system with the feedback of form (2.18) is

$$\dot{W}(x) = (A_1 - A_2)\omega_1\omega_2\omega_3 - h(x)\,(A_1^2\omega_1^2 + A_2^2\omega_2^2).$$

According to [5, Theorem 39.1], it is sufficient to show that

$$h(x)\,(A_1^2\omega_1^2 + A_2^2\omega_2^2) \geq (A_1 - A_2)\omega_1\omega_2\omega_3, \quad (h(x) > 0). \tag{2.26}$$

By using the inequality

$$2A_1A_2|\omega_1\omega_2| \leq A_1^2\omega_1^2 + A_2^2\omega_2^2,$$

we define $h(x)$ in the following manner:

$$h(x) = \left| \frac{A_1 - A_2}{2A_1A_2}\omega_3 \right| + \varepsilon, \tag{2.27}$$

where ε is an arbitrary positive constant. Then condition (2.26) holds.

Finally, by taking into account (2.24), (2.25), (2.27), expression (2.18) takes the form:

$$u_1 = \omega_2\omega_3 - \frac{\varkappa}{A_1}v_2v_3 - \left\{ \frac{|A_1 - A_2|}{2A_2}|\omega_3| + \varepsilon A_1 \right\}\omega_1,$$

$$u_2 = -\omega_1\omega_3 + \frac{\varkappa}{A_2}v_1v_3 - \left\{ \frac{|A_1 - A_2|}{2A_1}|\omega_3| + \varepsilon A_2 \right\}\omega_2, \quad (\varkappa > 0,\ \varepsilon > 0). \tag{2.28}$$

Let us remark that the feedback law (2.28) not only stabilizes the solution (2.21) of system (2.19) and (2.20) with respect to variables (2.23), but also ensures Lyapunov stability of the solution (2.21) due to the inequality $\dot{W} \leq 0$ (see Fig. 2.2).

Figure 2.2 illustrates the solution of the closed-loop system (2.19), (2.20), (2.28) for the following parameters[1]:

$$A_1 = 1, \quad A_2 = 3/2, \quad A_3 = 2, \quad \varkappa = 1, \quad \varepsilon = 1/10,$$

and initial conditions

$$\omega(0) = 0, \quad v_1(0) = 1/\sqrt{3}, \quad v_2(0) = v_3(0) = 0.$$

We see in Fig. 2.2 that the components $\omega_1(t)$, $\omega_2(t)$, $v_1(t)$, and $v_2(t)$ of this solution of the closed-loop system tend to zero for large t. We also note that the limit motion of the satellite corresponds to uniform rotations around v with constant angular velocity $\omega_3 \neq 0$.

---

[1] To simplify notations, we assume that all state variables and parameters are dimensionless in this section.

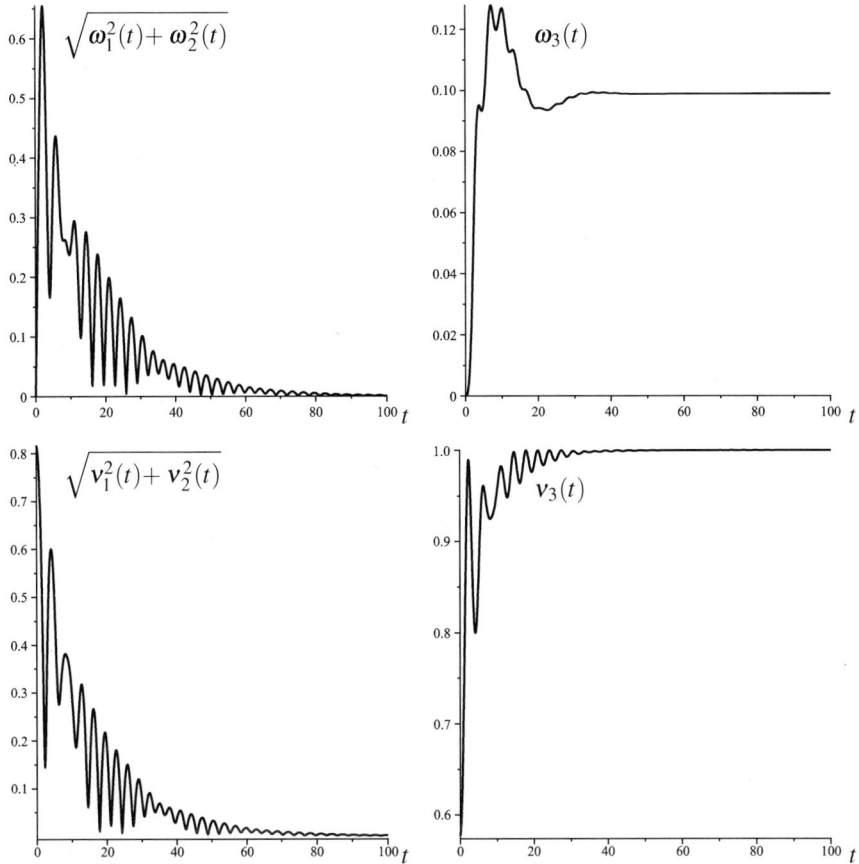

Fig. 2.2 Solution of the closed-loop system (2.19), (2.20) with control (2.28)

2.3.3 A Satellite with Moving Masses

Let us consider a system describing the rotation of a satellite with a pair of flywheels (see Fig. 2.3):

$$(A_1 - I_1)\dot{\omega}_1 = (A_2 - A_3)\omega_2\omega_3 + I_2\Omega_2\omega_3 - u_1,$$
$$(A_2 - I_2)\dot{\omega}_2 = (A_3 - A_1)\omega_1\omega_3 - I_1\Omega_1\omega_3 - u_2,$$
$$A_3\dot{\omega}_3 = (A_1 - A_2)\omega_1\omega_2 + I_1\Omega_1\omega_2 - I_2\Omega_2\omega_1,$$
$$I_1(\dot{\Omega}_1 + \dot{\omega}_1) = u_1, \quad I_2(\dot{\Omega}_2 + \dot{\omega}_2) = u_2,$$
$$\dot{v}_1 = \omega_3 v_2 - \omega_2 v_3, \quad \dot{v}_2 = \omega_1 v_3 - \omega_3 v_1, \quad \dot{v}_3 = \omega_2 v_1 - \omega_1 v_2. \quad (2.29)$$

Fig. 2.3 Satellite controlled
by a pair of flywheels

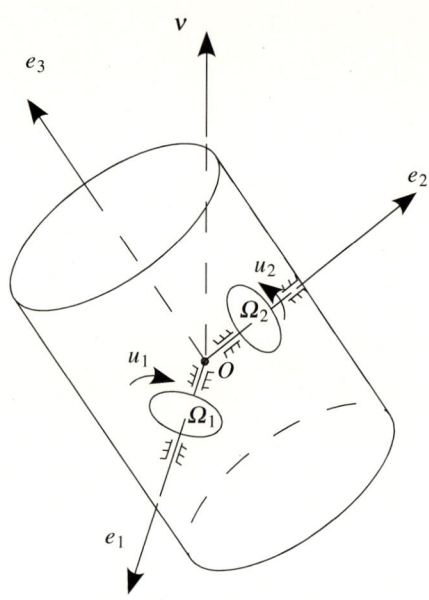

As in the previous example, the dynamics of the carrier body is characterized by the
angular velocity vector $\omega = \omega_1 e_1 + \omega_2 e_2 + \omega_2 e_2$ and coordinates of the fixed unit
vector $v = v_1 e_1 + v_2 e_2 + v_3 e_3$ (see, e.g., [16]). We assume that the ith flywheel
rotates around the direction of e_i with relative angular velocity Ω_i under the action
of control torque u_i, and the moment of inertia of the ith flywheels is denoted by I_i,
$i = 1, 2$ [15]. The control torques u_1 and u_2 are implemented by electric motors.
We denote by A_i the moment of inertia of the whole system (i.e. the carrier body
and flywheels) with respect to the e_i axis, $i = 1, 2, 3$. We assume that the motion
of flywheels does not change the mass distribution in the system, and that A_i are
principal moments of inertia.

System (2.29) admits the following equilibria for $u_1 = u_2 = 0$:

$$\omega = 0, \ \Omega_1 = \text{const}, \ \Omega_2 = \text{const}, \ v_1 = v_2 = 0, \ v_3 = 1. \tag{2.30}$$

The equilibrium (2.30) cannot be stabilized with respect to all variables, since system (2.29) admits the following integrals:

$$\Phi_1 = (A_1\omega_1 + I_1\Omega_1)^2 + (A_2\omega_2 + I_2\Omega_2)^2 + (A_3\omega_3)^2 = \text{const};$$
$$\Phi_2 = (A_1\omega_1 + I_1\Omega_1)v_1 + (A_2\omega_2 + I_2\Omega_2)v_2 + A_3\omega_3 v_3 = \text{const};$$
$$\Phi_3 = v_1^2 + v_2^2 + v_3^2 = \text{const}.$$

In order to stabilize solution (2.30) of system (2.29) with respect to variables $y = (\omega_1, \omega_2, v_1, v_2)$, we apply Theorem 2.3 with the following Lyapunov function:

$$2V(x) = (A_1 - I_1)\omega_1^2 + (A_2 - I_2)\omega_2^2 + \varkappa(v_1^2 + v_2^2), \quad \varkappa > 0.$$

Then

$$a(x) = (A_2 - A_1)\omega_1\omega_2\omega_3 + (I_2\Omega_2\omega_1 - I_1\Omega_1\omega_2)\omega_3 + \varkappa v_3(\omega_1 v_2 - \omega_2 v_1),$$
$$b_1(x) = -\omega_1, \ b_2(x) = -\omega_2.$$

The function $u^0(x)$ from condition (2) of Theorem 2.3 is a solution of the following algebraic equation:

$$\{(A_2\omega_2 + I_2\Omega_2)\omega_3 + \varkappa v_2 v_3 - u_1^0\}\omega_1 - \{(A_1\omega_1 + I_1\Omega_1)\omega_3 + \varkappa v_1 v_3 + u_2^0\}\omega_2 = 0.$$

To satisfy this equation, we assume

$$u_1^0 = \varkappa v_2 v_3 + (A_2\omega_2 + I_2\Omega_2)\omega_3, \ u_2^0 = -\varkappa v_1 v_3 - (A_1\omega_1 + I_1\Omega_1)\omega_3. \quad (2.31)$$

It can be seen that the set

$$M_1 = \{(\omega_1, \omega_2, \omega_3, \Omega_1, \Omega_2, v_1, v_2, v_3) \mid \omega_1 = \omega_2 = 0, \ v_1^2 + v_2^2 \neq 0\}$$

does not contain any positive semi-trajectory of system (2.29) with the feedback law $u = u^0(x)$ in a neighborhood of (2.30).

Let us assume $h(x) = \varepsilon = \text{const}$ and show that the feedback law

$$u_1 = \varkappa v_2 v_3 + (A_2\omega_2 + I_2\Omega_2)\omega_3 + \varepsilon\omega_1,$$
$$u_2 = -\varkappa v_1 v_3 - (A_1\omega_1 + I_1\Omega_1)\omega_3 + \varepsilon\omega_2. \quad (2.32)$$

satisfied condition (3) of Theorem 2.3 for any $\varkappa > 0$ and $\varepsilon > 0$. Indeed, the boundedness of the solutions with respect to variables $(\omega_3, \Omega_1, \Omega_2, v_3)$ follows from integrals Φ_1 and Φ_3 of system (2.29).

Thus, the solution (2.30) of system (2.29) is stabilizable with respect to variables (2.23) by means of the feedback law (2.32) by Theorem 2.3.

In order to illustrate the proposed stabilization scheme, we perform a numerical integration of the closed-loop system (2.29), (2.32) with the following parameters:

$$A_1 = 2, \ A_2 = 3, \ A_3 = 4, \ I_1 = I_2 = 1, \ \varepsilon = 1/10, \ \varkappa = 1.$$

Time-plots of the solution of the closed loop-system are shown in Figs. 2.4 and 2.5 for the initial conditions

$$\omega(0) = 0, \ \Omega_1(0) = \Omega_2(0) = 0, \ v_1(0) = v_3(0) = 0, \ v_2(0) = -1.$$

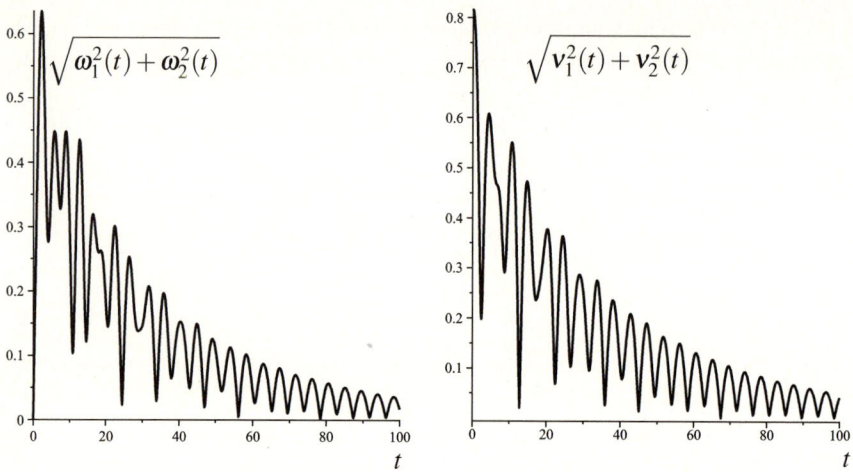

Fig. 2.4 Solution components of the closed-loop system (2.29) with control (2.32)

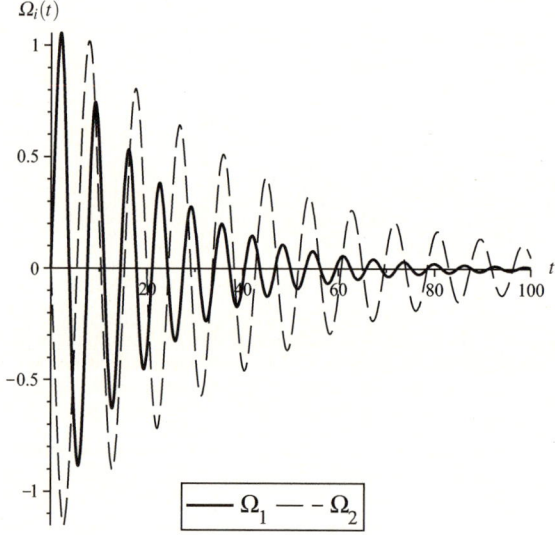

Fig. 2.5 Time plots of $\Omega_1(t)$ and $\Omega_2(2)$

Figures 2.4 and 2.5 confirm that the feedback law (2.32) stabilizes the system with respect to variables (2.23). We also observe that the limit position of the carrier body approximately corresponds to its equilibrium with $e_3 = v$, and the flywheels perform rotations with decaying angular velocities $\Omega_1(t)$ and $\Omega_2(t)$.

2.4 Partial Asymptotic Stability of Nonlinear Semigroups

In Sect. 2.1, we have obtained a general result on partial asymptotic stability without assuming the uniqueness of solutions as well as the differentiability of a Lyapunov functional. To derive more convenient stability conditions for the analysis of distributed parameter systems, let us consider a class of dynamical systems governed by differential equations in a Banach space.

Let E be a Banach space with the norm $\| \cdot \|$, and let X be its closed subset containing some ball $B_R = \{x \in E \mid \|x\| \leq R\}$ of radius $R > 0$. Then X is a metric space with respect to the distance $\rho(a, b) = \|a - b\|$. Let F be a (nonlinear) operator from $D(F) \subset X$ into E, F is supposed to be closed and densely defined on X. Given $x^0 \in X$ consider the abstract Cauchy problem (cf. [19, Chap. 4], [20, Sect. 5.2]) for F with initial data x^0:

$$\frac{dx(t)}{dt} = Fx(t), \quad t \in \mathbb{R}^+, \ x(0) = x^0. \tag{2.33}$$

We assume that the operator F generates a *nonlinear continuous semigroup* on X in the sense of Definition 1.9.

As F is the infinitesimal generator of a continuous semigroup $\{S(t)\}$, the Cauchy problem (2.33) is well-posed, and any mild solution of (2.33) is given by

$$x(t) = S(t)x^0, \quad t \in \mathbb{R}^+, \ x^0 \in X.$$

In that case, at each x^0 we may associate the singleton $\pi(x^0) = \{S(\cdot)x^0\}$. It is easy to check that the above defined $\pi : X \to 2^K$ is a multivalued D-system in the sense of Definition 2.1. (Assumption A_5 is satisfied by uniqueness of the solutions and A_4 is a consequence of continuity of the map $(t, x) \mapsto S(t)x$.)

Let $V : E \to \mathbb{R}$ be a differentiable functional, then $V(S(t)x^0)$ is differentiable along any classical solution of (2.33). The time-derivative of V along the trajectories of (2.33) at $x \in X$ is defined as

$$\dot{V}(x) = \lim_{t \to +0} \frac{V(S(t)x) - V(x)}{t}.$$

The above expression can also be written in the terms of vector fields for $x \in D(F)$:

$$\dot{V}(x) = (DV(x), Fx), \tag{2.34}$$

where $(\cdot, \cdot) : E^* \times E \to \mathbb{R}$ is the duality pairing of E and E^*, i.e. $(DV(x), \xi)$ is the value of a linear functional $DV(x) \in E^*$ at $\xi \in E$. When E is a Hilbert space, (2.34) takes the form

$$\dot{V}(x^0) = \langle \nabla V(x), Fx \rangle.$$

Here $\langle \cdot, \cdot \rangle$ is the scalar product in E, ∇ denotes the gradient.

As a consequence of Theorem 2.1 and the regularity of V, we have

Theorem 2.4 *Let F be the infinitesimal generator of a nonlinear continuous semigroup $\{S(t)\}$ on X, $F(0) = 0$, and let $\mu : X \to \mathbb{R}^+$ be a continuous functional. Assume that there exists a differentiable functional $V : E \to \mathbb{R}$ satisfying the following conditions.*

(i) *There exist $\alpha_1(\cdot), \alpha_2(\cdot) \in \mathcal{K}$ such that*

$$\alpha_1(\mu(x)) \le V(x) \le \alpha_2(\|x\|) \quad \text{for all } x \in X.$$

(ii) *$\dot{V}(x) \le 0$, for all $x \in D(F)$.*

(iii) *There exists $\Delta > 0$ such that*

$$\bigcup_{t \ge 0} \{S(t)x^0\}$$

is precompact in X provided that $\|x^0\| < \Delta$.

(iv) *$\operatorname{Ker}\mu = \{x \in X \mid \mu(x) = 0\}$ is invariant for (2.33), i.e. $\mu(S(\tau)x^0) = 0$ and $\tau \ge 0$ imply $\mu(S(t)x^0) = 0$ for all $t \in \mathbb{R}^+$.*

(v) *The set*

$$M = \overline{\{x \in D(F) \mid \dot{V}(x) = 0\}} \setminus \operatorname{Ker}\mu$$

does not contain any semitrajectory of (2.33) defined for $t \in \mathbb{R}^+$.

Then the equilibrium $x^0 = 0$ of (2.33) is asymptotically stable with respect to μ.

Proof As $F(0) = 0$, the solution $x(t) \equiv 0$ is an equilibrium of the multivalued D-system defined by $x^0 \in X \mapsto \pi(x^0) = \{S(\cdot)x^0\}$. It is easy to see that (i) and (iii) imply C_1 and C_3 in conditions of Theorem 2.1.

Let us prove that condition (ii) implies that $V(x(t))$ is non-increasing on any mild solution of (2.33) with $x^0 \in X$, $t \in \mathbb{R}^+$. If $x^0 \in D(F)$ then $x(t) = S(t)x^0$ is a classical solution, and $\dot{V}(x(t))$ given by formula (2.34) exists for all $t \ge 0$. As $V(x(t))$ is continuous and $\dot{V}(x(t)) \le 0$ on $\mathbb{R}^+$, the function $V(x(t))$ is non-increasing on $\mathbb{R}^+$. For arbitrary $x^0 \in X \setminus D(F)$ and $T > 0$, the mild solution $S(t)x^0$ ($0 \le t \le T$) can be approximated by classical ones in the $L^\infty([0, T]; E)$ norm (it is a consequence of assumption A_4). Therefore, as V is non-increasing along any classical solution and V is continuous, $V(S(t)x^0)$ is non-increasing on $\mathbb{R}^+$ for each $x^0 \in X$.

To finish the proof, let us show that C_4 holds under our assumptions. If $p \in M_1$ in (2.5) then $\frac{d}{dt}V(S(t)p) = 0$ for all $t \in \mathbb{R}^+$. Therefore,

$$M_1 \subset M_0 = \overline{\{x \in D(F) \mid \dot{V}(x) = 0\}}.$$

(The closure in M_0 is taken because (2.34) defines $\dot{V}$ on $D(F)$ only, but F is densely defined.) On the other hand, as M_1 is semi-invariant,

$$M_1 \subset \{x \in M_0 \mid S(t)x \in M_0 \text{ for all } t \in \mathbb{R}^+\}. \tag{2.35}$$

Suppose that x be an element of the right-hand side of (2.35). The kernel invariance (iv) implies either $S(t)x \in \operatorname{Ker} \mu$ or $S(t)x \notin \operatorname{Ker} \mu$ for all $t \in \mathbb{R}^+$. But the last case is impossible because of (v). Therefore, any $x \in M_1$ belongs to $\operatorname{Ker} \mu$ that proves C_4. $\square$

Remark 2.6 The above theorem can be applied for studying strong asymptotic stability when $\mu(x) = \|x\|$. If F is linear then Definition 1.9 is equivalent to that of C_0-semigroups of linear bounded operators. Therefore, the assumption of Theorem 2.4 regarding the semigroup $\{S(t)\}$ can be checked via the Hille–Yosida or the Lumer–Phillips theorems for linear operators F (cf. [19, Chap. 1]). In more general case, there is a close relationship between quasicontractive semigroups (which are jointly continuous) and ω-accretive operators F [21]. If $\omega = 0$ then the generator of $\{S(t)\}$ is dissipative (cf. [22, Sect. 2.9]). The compactness assumption (iii) can be checked by the method of [23] for a class of monotone operators. We extend this approach for the case of bounded perturbations of C_0-semigroups in the next section.

2.5 Relative Compactness of Trajectories in a Banach Space

In order to apply Theorem 2.4, it is necessary to check the relative compactness of trajectories for a differential equation in a Banach space. One could observe that the accretivity condition (monotonicity), which is a crucial assumption of the paper [23], is violated for some important classes of flexible systems. In particular, the infinitesimal generator of the nonlinear system considered in [24] is not monotone. This fact stimulates the development of new tools for the analysis of the compactness for trajectories of distributed parameter systems. This section provides some compactness results based on a priori estimates of perturbations for a differential equation in a Banach space.

Let E be a real Banach space, and let $A : D(A) \to E$ be a closed linear operator with the domain of definition of $D(A) \subset E$. Consider the abstract Cauchy problem for $t \in [0, +\infty)$:

$$\dot{x}(t) = Ax(t), \quad x(0) = x_0 \in E. \tag{2.36}$$

We assume that the domain $D(A)$ is dense in E, and that A is the infinitesimal generator of a C_0-semigroup of linear operators $\{e^{tA}\}_{t \geq 0}$ in E. So, the Cauchy problem (2.36) is well posed for $t \in [0, +\infty)$, and its mild solution can be represented in the form

$$x(t) = e^{tA}x_0, \quad t \geq 0. \tag{2.37}$$

In order to study a wider class of equations (including the one with non-monotone operators), we introduce the perturbed Cauchy problem for $t \geq 0$ as follows:

$$\dot{x} = Ax + f(t)R(x, t), \quad x(0) = x_0 \in E, \tag{2.38}$$

where $f : [0, +\infty) \to \mathbb{R}$ and $R : E \times [0, +\infty) \to E$ are continuous mappings.

We prove that the compactness property is preserved by passing from Eqs. (2.36)–(2.38) under some additional assumptions on the function f and the mapping R. This result will be applied to derive sufficient conditions for the compactness of trajectories of an autonomous differential equation in a Banach space.

2.5.1 Compactness Lemmas

Assume that the Banach space E has a basis $\{e_i\}$ $(i = 1, 2, \ldots)$. We denote by $\{f_j\} \subset E^*$ $(j = 1, 2, \ldots)$ the conjugate system of bounded linear functionals, i.e. $f_j(e_i) = \delta_{ij}$, where δ_{ij} is the Kronecker delta. Then, for each $x \in E, n \in \mathbb{N}$, we define the linear projection operators:

$$S_n(x) = \sum_{i=1}^{n} f_i(x)e_i, \quad P_n(x) = x - S_n(x).$$

As $\{e_i\}$ is a basis then the operators $S_n : E \to E$ are uniformly bounded:

$$\|S_n\| \leq M < \infty, \quad n = 1, 2, \ldots$$

To describe the compact subsets of E, we formulate two auxiliary results.

Lemma 2.4 *Let $\{e_i\}$ be a basis of E. A bounded subset $C \subset E$ is relatively compact in E iff*

$$\lim_{n \to \infty} \sup_{x \in C} \|P_n x\| = 0. \tag{2.39}$$

Proof If C is precompact then the Hausdorff compactness criterion implies that, for any $\varepsilon > 0$, there exists a finite $\frac{\varepsilon}{2M}$-net $\{x^{(j)}\}$, $j = 1, 2, \ldots, m(\varepsilon)$ [25]. It means that, for each $x \in C$, there is a $j \leq m(\varepsilon)$ such that $\|x - x^{(j)}\| < \frac{\varepsilon}{2M}$, i.e.

$$\|P_n x\| = \|P_n(x - x^{(j)}) + P_n x^{(j)}\| \leq \|P_n\|\frac{\varepsilon}{2M} + \|P_n x^{(j)}\| < \frac{\varepsilon}{2} + \|P_n x^{(j)}\|. \tag{2.40}$$

We now show that, for sufficiently large n, the inequality $\|P_n x^{(j)}\| < \frac{\varepsilon}{2}$ holds for each $j = 1, 2, \ldots, m(\varepsilon)$. In fact, since $\{e_i\}$ is a basis, each element of the net is represented by a convergent series:

$$x^{(j)} = \sum_{i=1}^{\infty} c_i^{(j)} e_i. \tag{2.41}$$

According to the definition of P_n,

$$\| P_n x^{(j)} \| = \| \sum_{i=n+1}^{\infty} c_i^{(j)} e_i \|.$$

The last expression does not exceed $\frac{\varepsilon}{2}$, starting with some index $n = n(\varepsilon)$, since the series in (2.41) is convergent in the norm of E. Thus, (2.40) implies $\| P_n x \| < \varepsilon$ for $n \geq n(\varepsilon)$, which proves (2.39).

Conversely, if the set C is bounded, then all finite-dimensional projection

$$C_n = \{ S_n x \mid x \in C \}, \quad n = 1, 2, \ldots$$

are precompact. Relation (2.39) implies that, for each $\varepsilon > 0$, any finite ε-net of the set C_n can be used to cover C for sufficiently large n.

We will call a C_0-semigroup of linear operators $\{ e^{tA} \}_{t \geq 0}$ in E *uniformly bounded* if [19]:

$$\| e^{tA} \| \leq N, \quad \forall t \geq 0,$$

with some constant $N < \infty$.

Lemma 2.5 *Assume that $\{ e_i \}$ is a basis in E, C is a compact subset of E, and $\{ e^{tA} \}_{t \geq 0}$ is a uniformly bounded C_0-semigroup of linear operators in E, for which the trajectory $\gamma(x_0) = \{ e^{tA} x_0 \mid t \geq 0 \}$ is precompact for any $x_0 \in C$. Then*

$$\lim_{n \to \infty} \left(\sup_{t \geq 0, x \in C} \| P_n e^{tA} x \| \right) = 0. \tag{2.42}$$

Proof According to Lemma 2.4, to prove (2.42) is suffices to establish that the set

$$K = \{ e^{tA} \mid x \in C, t \geq 0 \}.$$

is precompact. Let $\{ y_n \}$ be a sequence of elements of K, i.e. $y_n = e^{t_n A} x_n$ for some $\{ t_n \} \subset [0, +\infty)$, $\{ x_n \} \subset C$, $n = 1, 2, \ldots$ The compactness of C implies the existence of a convergent subsequence $x_{n(k)} \to x^* \in C$ as $k \to \infty$. As $\gamma(x^*)$ is precompact then there exists a convergent subsequence $e^{t_{n(k(m))} A} x^* \to y^* \in E$ as $m \to \infty$. By using the uniformly bounded semigroup $\{ e^{tA} \}_{t \geq 0}$, we conclude that $e^{t_{n(k(m))} A} x_{n(k(m))} \to y^*$ as $m \to \infty$.

2.5.2 Trajectories of the Perturbed System

Let us recall that *a mild solution* of the inhomogeneous problem (2.38) on $0 \leq t < T \leq +\infty$ is a continuous function $x : [0, T) \to E$ that satisfies the integral equation [19]:

$$x(t) = e^{tA}x_0 + \int_0^t e^{(t-s)A} f(s) R(x(s), s) \, ds. \tag{2.43}$$

The integral in formula (2.43) in treated the sense of Bochner. We formulate the following sufficient condition for the precompactness of trajectories to the perturbed differential equation.

Theorem 2.5 *Let E be a Banach space with a basis, A be the infinitesimal generator of a uniformly bounded C_0-semigroup of linear operators $\{e^{tA}\}_{t\geq0}$ in E, $f \in L^1[0, +\infty)$, $R(x, t) \in K$ for all $x \in E$, $t \geq 0$, and K is a compact set. Assume that the set $\{e^{tA}y \mid t \geq 0\}$ is precompact for all $y \in K \cup \{x_0\}$.*

Then each mild trajectory $\{x(t) \mid t \geq 0\}$ of (2.38) is contained in a compact subset of E.

Proof Let $x(t)$ be a mild solution of (2.38) on the semi-interval $t \geq 0$. Then the integral equation (2.43) implies the compactness of $\{e^{tA}x_0 \mid t \geq 0\}$, and condition $f \in L^1[0, +\infty)$ together with $R \in K$ provides the boundedness of $x(t)$. According to Lemma 2.4, to prove the precompactness of $\{x(t) \mid t \geq 0\}$ is suffices to choose a basis $\{e_i\}$ in E and establish the existence of the limit

$$\lim_{n\to\infty} \sup_{t\geq0} \|P_n x(t)\| = 0.$$

Applying the projection operator to (2.43), we get

$$\|P_n x(t)\| \leq \|P_n e^{tA}x_0\| + \left\| \int_0^t f(s) P_n \left(e^{(t-s)A} R(x(s), s) \right) ds \right\|$$

$$\leq \|P_n e^{tA}x_0\| + \|f\|_{L^1} \cdot \sup_{s\in[0,t], y\in K} \|P_n e^{sA}y\|.$$

The proof is completed by applying Lemmas 2.4 and 2.5.

   A certain class of autonomous differential equations with nonlinear infinitesimal generators can be transformed to the form (2.38) with an appropriate assumption on the function $f(t)$. We state the main result in this direction for the following abstract Cauchy problem:

$$\dot{x}(t) = Ax(t) + h(x(t))B(x(t)), \quad x(0) = x_0 \in E, \tag{2.44}$$

where $h : E \to \mathbb{R}$ and $B : E \to E$ are locally Lipschitz mappings. Recall that the mappings $h(x)$ and $B(x)$ are called *locally Lipschitz* if, for every $r \geq 0$, there exists a constant $L(r)$ such that

$$|h(x) - h(y)| \leq L(r)\|x - y\|, \quad \|B(x) - B(y)\| \leq L(r)\|x - y\|$$

for all $\|x\| \leq r$, $\|y\| \leq r$. If $w : E \to \mathbb{R}$ is a Fréchet differentiable functional, then the function of time $w(x(t))$ is differentiable along each classical solution $x(t)$ to the problem (2.44). Then, for any $x \in D(A) \subset E$, the time-derivative of w along the trajectories of (2.44) can be written as

$$\dot{w}(x) = (Ax + B(x)h(x), \nabla_x w),$$

where $(\cdot, \cdot) : E \times E^* \to \mathbb{R}$ is the duality pairing of E and E^*, i.e. $(\xi, \nabla_x w)$ is the value of the linear functional $\nabla_x w \in E^*$ at $\xi \in E$.

Theorem 2.6 *Assume that E is a Banach space with a basis, A is the infinitesimal generator of a uniformly bounded C_0-semigroup of linear operators $\{e^{tA}\}_{t\geq 0}$ in E, the set $\{e^{tA} y \mid t \geq 0\}$ is precompact for all $y \in E$, and $B : E \to E$ is a compact operator. Assume, moreover, that $w : E \to \mathbb{R}$ is a Fréchet differentiable functional that satisfies the following conditions:*

(1) *the set $M_c = \{x \mid w(x) \leq c\}$ is bounded for each $c \in \mathbb{R}$;*
(2) *$\inf_{\|x\| \leq r} w(x) > -\infty$ for all $r > 0$;*
(3) *there exists a constant $k_1 > 0$ such that*

$$\dot{w}(x) \leq k_1 h(x) \leq 0, \quad \forall x \in D(A).$$

Then, for each $x_0 \in E$, the Cauchy problem (2.44) has the unique solution $x(t)$ on $[0, +\infty)$, and $\{x(t) \mid t \geq 0\}$ is precompact in E.

Proof According to Theorem 1.4 [19], for each $x_0 \in E$, there exists a unique maximal mild solution $x(t)$ of the problem (2.44), $t \in [0, t_{max})$. Conditions (1) and (3) imply that $x(t)$ is bounded, hence, $t_{max} = +\infty$. Let us consider equation (2.38) with $R(x, t) = B(x)$ and $f(t) = h(x(t))$. Then conditions (2) and (3) yield the property $f \in L^1[0, +\infty)$. Thus, the trajectory $\{x(t) \mid t \geq 0\}$ is precompact in E by Theorem 2.5.

Note that, since the set M_c is forward invariant under the condition $\dot{w}(x) \leq 0$, then Theorem 2.6 admits a local formulation on the subset of E located between level surfaces of the functional w.

References

1. Zuyev, A.: Partial asymptotic stability and stabilization of nonlinear abstract differential equations. In: Proceedings of the 42nd IEEE Conference on Decision and Control CDC'03, pp. 1321–1326. Maui, Hawaii (2003)
2. Rouche, N., Habets, P., Laloy, M.: Stability Theory by Liapunov's Direct Method. Springer, New York (1977)
3. LaSalle, J.P.: Stability theory and invariance principles. In: Cesari, L., Hale, J.K., LaSalle, J.P. (eds.) Dynamical Systems. Volume 1: International Symposium on Dynamics System Providence 1974, pp. 211–222. Academic Press, New York (1976)

4. Movčan, A.A.: Stability of processes with respect to two metrics. J. Appl. Math. Mech. **24**, 1506–1524 (1961)
5. Rumyantsev, V.V., Oziraner, A.S.: Stability and Stabilization of Motion with Respect to Part of Variables (in Russian). Nauka, Moscow (1987)
6. Vorotnikov, V.I.: Partial Stability and Control. Birkhäuser, Boston (1998)
7. Filippov, A.F.: Differential Equations with Discontinuous Righthand Sides. Kluwer, Dordrecht (1988)
8. Aubin, J.-P., Cellina, A.: Differential Inclusions. Springer, Berlin (1984)
9. Lyapunov, A.M.: The General Problem of the Stability of Motion (A.T. Fuller trans.) Taylor & Francis, London (1992)
10. Smirnov, G.V.: Introduction to the Theory of Differential Inclusions. AMS, Providence (2002)
11. Risito, C.: Sulla stabilita asintotica parziale. Ann. Math. Pura Appl. **84**, 279–292 (1970)
12. Krasovskii, N.N.: Problems of the Theory of Stability of Motion. Stanford University Press, Stanford (1963)
13. Kovalev, A.M.: Invariance and asymptotic stability. J. Appl. Math. Mech. **75**, 317–322 (2011)
14. Bacciotti, A., Ceragioli, F.: Stability and stabilization of discontinuous systems and nonsmooth Lyapunov functions. ESAIM: Control Optim. Calc. Var. **4**, 361–376 (1999)
15. Zuyev, A.: On partial stabilization of nonlinear autonomous systems: sufficient conditions and examples. In: Proceedings of the European Control Conference ECC'01, Porto, Portugal, pp. 1918–1922 (2001)
16. Zubov, V.I.: Lectures in Control Theory (in Russian). Nauka, Moscow (1975)
17. Agrachev, A.A., Sachkov, Yu.L.: Control Theory From the Geometric Viewpoint. Springer, Berlin (2004)
18. Sontag, E.D.: Mathematical Control Theory: Deterministic Finite Dimensional Systems, 2nd edn. Springer, New York (1998)
19. Pazy, A.: Semigroups of Linear Operators and Applications to Partial Differential Equations. Springer, New York (1983)
20. Fattorini, H.O.: Infinite Dimensional Optimization and Control Theory. Cambridge University Press, Cambridge (1999)
21. Barbu, V.: Analysis and Control of Nonlinear Infinite Dimensional Systems. Academic Press, San Diego (1992)
22. Luo, Z.-H., Guo, B.-Z., Morgul, O.: Stability and Stabilization of Infinite Dimensional Systems with Applications. Springer, London (1999)
23. Dafermos, C.M., Slemrod, M.: Asymptotic behavior of nonlinear contraction semigroups. J. Funct. Anal. **13**, 97–106 (1973)
24. Coron, J.-M., d'Andrea-Novel, B.: Stabilization of a rotating body beam without damping. IEEE Trans. Autom. Control **44**, 608–618 (1998)
25. Kantorovich, L.V., Akilov, G.P.: Functional Analysis. Pergamon Press, New York (1982)

Chapter 3
Stabilization of a Rotating Body with Euler–Bernoulli Beams

Abstract This chapter is focused on the partial stabilization problem of a rotating rigid body endowed with a number of elastic beams. To stabilize the equilibrium of this mechanical system, we apply results of Chap. 2. In addition, we prove strong (non-asymptotic) stability in the sense of Lyapunov as well as relative compactness of the trajectories for the corresponding nonlinear semigroup. In this chapter, we also study a mathematical model of a multi-link manipulator consisting of rigid bodies and a chain of Euler–Bernoulli beams.

Networks of elastic strings, thermoelastic beams, interconnected membranes, liked plates, and plate-beams systems have been studied in the monograph [1] by Lagnese, Leugering, and Schmidt. This monograph focuses on the analysis of the transient behavior of solutions and questions of well-posedness and stability, spectral properties and exact, spectral controllability of the transients.

A model of a rotating body with a flexible beam has been considered by Baillieul and Levi [2, 3], Bloch and Titi [4], Cai et al. [5, 6], Cai and Lim [7, 8], Laousy et al. [9], Xu and Baillieul [10]. The stabilization problem has been solved in the paper by Coron and d'Andrea-Novel [11] for a nonlinear version of such model without damping. It should be noted that the stabilization of this model faces significant technical difficulties as the rotation of the mechanical system is unstable in the absence of control. Thus, it is of great theoretical interest to stabilize systems that may be inherently unstable or that may posses only weak natural damping. For this purpose, we will consider models of flexible structures without natural damping in this book.

Problems of the optimal damping of vibrations have been addressed in the book by Krabs [12] for one-dimensional media such as strings and beams within the framework of the moments theory. In particular, the existence and uniqueness results have been obtained for problems of null-controllability and time-minimal null-controllability. We also refer the reader to papers [13–18] for a survey of other results concerning the modeling and control of flexible beams.

In contrast to known results in the literature, we focus here on the partial stabilization problem for a mechanical system that consists of a rotating rigid body with an arbitrary number of flexible beams.

© Springer International Publishing Switzerland 2015   39
A.L. Zuyev, *Partial Stabilization and Control of Distributed Parameter Systems with Elastic Elements*, Lecture Notes in Control and Information Sciences 458, DOI 10.1007/978-3-319-11532-0_3

3.1 Equations of Motion of a Rigid Body
   with Elastic Attachments

Consider a planar mechanical system consisting of a rigid body and k elastic beams (Fig. 3.1). This system is an approximate model of a controlled satellite with flexible antennae performing planar maneuvers (see, e.g., [1, 19, 20] for references and other models).

The rigid body can rotate around the fixed point O under the action of a control torque M. We suppose that each beam is attached at the distance d from the point O, and that l_i is the length of the ith beam. Let $w_i(x, t)$ be the deflection of the ith beam from the axis $O_i x_i$ at the location $x \in [0, l_i]$ and time $t \geq 0$. Thus, the evolution of the system is defined by the following functions:

$$\theta(t), \ w_i(x, t), \quad x \in [0, l_i], \ t \geq 0, \ i = \overline{1, k},$$

where θ is the angle between the axes Ox_0 and Ox_1.

The mechanical system, shown in Fig. 3.1, is a generalization of the model considered in [21] for the case of beams with different lengths.

We assume here that each beam is fixed at one of its end while the other end is free. This assumption implies the following boundary conditions on w_i (see, e.g., [22]):

$$w_i\Big|_{x=0} = \frac{\partial w_i}{\partial x}\Big|_{x=0} = \frac{\partial^2 w_i}{\partial x^2}\Big|_{x=l_i} = \frac{\partial^3 w_i}{\partial x^3}\Big|_{x=l_i} = 0, \quad i = \overline{1, k}. \tag{3.1}$$

Fig. 3.1 Rigid body with beams

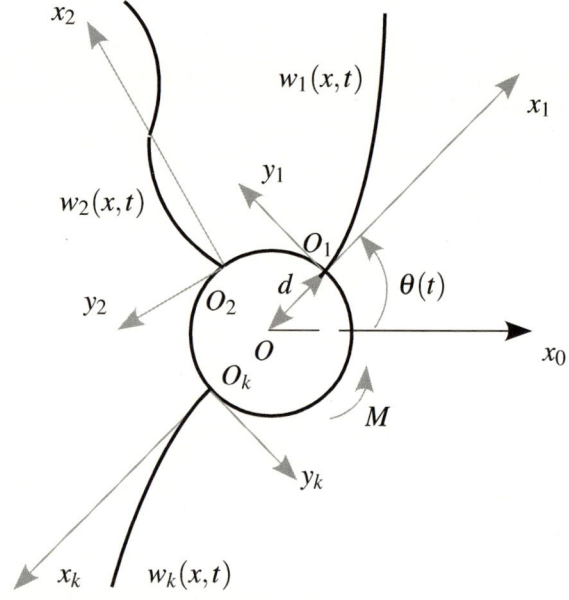

3.1.1 Hamilton's Principle

To derive the equations of motion, we consider the kinetic energy T of the system and the potential of elastic forces U by using the Euler–Bernoulli beam model (cf. [22]):

$$2T = J_0 \dot\theta^2 + \sum_{i=1}^{k} \int_0^{l_i} \left\{ (\dot\theta w_i)^2 + \left(\frac{\partial w_i}{\partial t} + \dot\theta(x + d) \right)^2 \right\} \rho_i \, dx, \qquad (3.2)$$

$$2U = \sum_{i=1}^{k} \int_0^{l_i} E_i I_i \left(\frac{\partial^2 w_i}{\partial x^2} \right)^2 dx, \qquad (3.3)$$

where $\dot\theta = \frac{d}{dt}\theta$, $J_0 > 0$ is the moment of inertia of the rigid body, and ρ_i is the mass per unit length of the ith beam. Values I_i and E_i denote the cross-sectional moment of inertia and Young's modulus of the ith beam, respectively. In the sequel, we assume that $E_i I_i$ and ρ_i are positive constants.

The Lagrangian of the system considered is defined as follows

$$L\left(\theta, \dot\theta, w_1(\cdot, t), \frac{\partial w_1(\cdot, t)}{\partial t}, \ldots, w_k(\cdot, t), \frac{\partial w_k(\cdot, t)}{\partial t} \right) = T - U.$$

Suppose that $\theta(t), w_1(x, t), \ldots, w_k(x, t)$ define the motion of the system for given $t \in [t_1, t_2]$ and control torque $M(t)$. Then Hamilton's principle [23, 24] yields

$$\delta \left(\int_{t_1}^{t_2} L \, dt \right) + \int_{t_1}^{t_2} M(t) \delta\theta(t) \, dt = 0, \qquad (3.4)$$

for each variation $\delta\theta(t), \delta w_i(x, t)$ such that

$$\delta\theta \in C^2[t_1, t_2], \quad \delta\theta(t_1) = \delta\theta(t_2) = 0,$$

$$\delta w_i \in C^2 \left([0, l_i] \times [t_1, t_2] \right), \quad \delta w_i(x, t_1) = \delta w_i(x, t_2) = 0, \quad \forall x \in [0, l_i],$$

$$\delta w_i|_{x=0} = \left. \frac{\partial(\delta w_i)}{\partial x} \right|_{x=0} = 0, \quad \forall t \in [t_1, t_2], \ i = 1, 2, \ldots, k. \qquad (3.5)$$

The variation of $\int_{t_1}^{t_2} L \, dt$ in (3.4) is obtained by computing the linear part of

$$\int_{t_1}^{t_2} \left\{ L\left(\theta + \delta\theta, \dot\theta + \delta\dot\theta, \ldots, w_k + \delta w_k, \frac{\partial}{\partial t}(w_k + \delta w_k) \right) - L\left(\theta, \dot\theta, \ldots, w_k, \frac{\partial}{\partial t} w_k \right) \right\} dt$$

and performing the integration by parts:

$$
\delta\left(\int_{t_1}^{t_2} L\,dt\right) + \int_{t_1}^{t_2} M\delta\theta\,dt
$$

$$
= \left\{ J_0\dot\theta + \sum_{i=1}^{k}\int_0^{l_i}\left(\dot\theta w_i{}^2 + (x+d)\frac{\partial w_i}{\partial t} + (x+d)^2\dot\theta\right)\rho_i dx\right\}\delta\theta(t)\Bigg|_{t=t_1}^{t_2}
$$

$$
+ \int_{t_1}^{t_2}\left\{ M - \frac{d}{dt}\left(J_0\dot\theta + \sum_{i=1}^{k}\int_0^{l_i}\left(\dot\theta w_i{}^2 + (x+d)\frac{\partial w_i}{\partial t} + (x+d)^2\dot\theta\right)\rho_i dx\right)\right\}\delta\theta\,dt
$$

$$
+ \sum_{i=1}^{k}\int_{t_1}^{t_2}\int_0^{l_i}\left\{\dot\theta^2 w_i - \frac{\partial}{\partial t}\left(\frac{\partial w_i}{\partial t} + (x+d)\dot\theta\right)\right\}\delta w_i(x,t)\rho_i\,dxdt
$$

$$
+ \sum_{i=1}^{k}\int_0^{l_i}\left\{\left(\frac{\partial w_i}{\partial t} + (x+d)\dot\theta\right)\delta w_i(x,t)|_{t=t_1}^{t_2}\right\}\rho_i dx
$$

$$
- \sum_{i=1}^{k}\int_{t_1}^{t_2}\left\{ E_i I_i \frac{\partial^2 w_i}{\partial x^2}\frac{\partial(\delta w_i)}{\partial x}\bigg|_{x=0}^{l_i}\right\}dt + \sum_{i=1}^{k}\int_{t_1}^{t_2}\int_0^{l_i}\left(E_i I_i \frac{\partial^3 w_i}{\partial x^3}\frac{\partial(\delta w_i)}{\partial x}\right)dx\,dt
$$

$$
= \int_{t_1}^{t_2}\left\{ M - \frac{d}{dt}\left(J_0\dot\theta + \sum_{i=1}^{k}\int_0^{l_i}\left(\dot\theta w_i{}^2 + (x+d)\frac{\partial w_i}{\partial t} + (x+d)^2\dot\theta\right)\rho_i dx\right)\right\}\delta\theta\,dt
$$

$$
+ \sum_{i=1}^{k}\int_{t_1}^{t_2}\int_0^{l_i}\left\{\dot\theta^2 w_i - \frac{\partial}{\partial t}\left(\frac{\partial w_i}{\partial t} + (x+d)\dot\theta\right) - c_i^2\frac{\partial^4 w_i}{\partial x^4}\right\}\delta w_i(x,t)\rho_i\,dxdt, \quad (3.6)
$$

where

$$
c_i = \sqrt{\frac{E_i I_i}{\rho_i}} > 0.
$$

We have used the fact that w_i and $(\delta\theta, \delta w_i)$ satisfy boundary conditions (3.1) and (3.5), respectively. As (3.6) vanishes on any admissible variation $\delta\theta, \delta w_1, \ldots, \delta w_k$, we have

$$
\frac{d}{dt}\left(J_0\dot\theta + \sum_{i=1}^{k}\int_0^{l}\left(\dot\theta w_i{}^2 + (x+d)\frac{\partial w_i}{\partial t} + (x+d)^2\dot\theta\right)\rho dx\right) = M;
$$

$$
\frac{\partial^2 w_i}{\partial t^2} + c^2\frac{\partial^4 w_i}{\partial x^4} + (x+d)\ddot\theta - \dot\theta^2 w_i = 0, \quad t \in [t_1,t_2],\ x \in [0,l],\ i = 1,2,\ldots,k.
$$

By rewriting the above system with respect to second order time-derivatives, we get

$$\ddot{\theta} = \left(J_0 + \sum_{i=1}^{k} \int_0^{l_i} w_i{}^2 \rho_i dx \right)^{-1} \left(M + \sum_{i=1}^{k} \int_0^{l_i} \left\{ (x+d)(c_i^2 \frac{\partial^4 w_i}{\partial x^4} - \dot{\theta}^2 w_i) - 2\dot{\theta} \frac{\partial w_i}{\partial t} w_i \right\} \rho_i dx \right);$$

$$\frac{\partial^2 w_j}{\partial t^2} = -c_i^2 \frac{\partial^4 w_j}{\partial x^4} + \dot{\theta}^2 w_j - (x+d) \left(J_0 + \sum_{i=1}^{k} \int_0^{l_i} w_i{}^2 \rho_i dx \right)^{-1}$$

$$\times \left(M + \sum_{i=1}^{k} \int_0^{l_i} \left\{ (x+d)(c_i^2 \frac{\partial^4 w_i}{\partial x^4} - \dot{\theta}^2 w_i) - 2\dot{\theta} \frac{\partial w_i}{\partial t} w_i \right\} \rho_i dx \right), \quad j = 1, 2, \ldots, k.$$

To simplify these equations, we denote

$$v = \frac{M + \sum_{i=1}^{k} \int_0^{l_i} \left\{ (x+d)(c_i^2 \frac{\partial^4 w_i}{\partial x^4} - \dot{\theta}^2 w_i) - 2\dot{\theta} \frac{\partial w_i}{\partial t} w_i \right\} \rho_i dx}{J_0 + \sum_{i=1}^{k} \int_0^{l_i} w_i{}^2 \rho_i dx} \qquad (3.7)$$

and treat v as a new control. Note that the denominator in formula (3.7) is positive as $J_0 > 0$. Given the state $(\theta, \dot{\theta}, \ldots, w_k(\cdot, t), \frac{\partial w_k(\cdot, t)}{\partial t})$ and the control v, the torque M can be easily reconstructed from the feedback transformation (3.7). Thus, the dynamics is described by the following differential equations

$$\ddot{\theta}(t) = v, \quad v \in \mathbb{R},$$

$$\frac{\partial^2 w_i(x, t)}{\partial t^2} + c_i^2 \frac{\partial^4 w_i(x, t)}{\partial x^4} = \dot{\theta}^2(t) w_i(x, t) \qquad (3.8)$$

$$- (x+d)v, \quad t \geq 0, \ x \in [0, l_i], \ i = 1, 2, \ldots, k,$$

and boundary conditions (3.1). Our goal is to study the stabilizability of the equilibrium $\theta = 0$, $w_i = 0$ for control system (3.1), (3.8) by means of a state feedback law.

3.1.2 Equations with Elastic Coordinates

In this subsection, we expand the transverse displacement $w_i(x, t)$ into the Fourier series

$$w_i(x, t) = \sum_{n=1}^{\infty} u_{in}(x) q_{in}(t) \qquad (3.9)$$

and treat the coefficients $q_{in}(t)$ as generalized coordinates. For each beam number $i \in \{1, 2, \ldots, k\}$, we define $u_{in}(x)$ as eigenfunctions of the following spectral problem:

$$\frac{d^4}{dx^4} u_{in}(x) = \lambda_{in} u_{in}(x), \quad x \in (0, l_i),$$ (3.10)

$$u_{in}(0) = u'_{in}(0) = u''_{in}(l_i) = u'''_{in}(l_i) = 0.$$ (3.11)

It is well-known that, for each fixed $i \in \{1, 2, \ldots, k\}$, the above problem has a countable set of eigenvalues $\lambda_{in} \in \mathbb{R}$ with properties

$$0 < \lambda_{i1} < \lambda_{i2} < \cdots < \lambda_{in} < \cdots, \quad \lambda_{in} = O(n^4) \text{ as } n \to \infty,$$ (3.12)

and the corresponding eigenfunctions $\{u_{in}(x)\}_{n=1}^{\infty}$ form an orthogonal basis in $L^2(0, l_i)$ (see, e.g., [25, p. 176]).

Thus, any transverse displacement $w_i(x, t)$ such that $w_i(\cdot, t) \in L^2(0, l_i)$ for fixed t, is represented in the form (3.9) uniquely.

To study properties of solutions of (3.10) and (3.11), we introduce the standard fourth-order spectral problem on $x \in [0, 1]$:

$$\frac{d^4}{dx^4} \phi_n(x) = \beta_n^4 \phi_n(x), \quad x \in (0, 1),$$

$$\phi_n(0) = \phi'_n(0) = \phi''_n(1) = \phi'''_n(1) = 0.$$

Let β_n^4 and $\phi_n(x)$ be, respectively, the eigenvalues and eigenfunctions of this problem for $n = 1, 2, \ldots$, Then β_n satisfy the transcendent equation

$$1 + \cos(\beta_n) \cosh(\beta_n) = 0,$$ (3.13)

and functions ϕ_n take the following form, up to a constant factor (cf. [25, p. 176]):

$$\phi_n(x) = -\frac{1 + \gamma_n}{2} e^{\beta_n x} - \frac{1 - \gamma_n}{2} e^{-\beta_n x} + \gamma_n \sin(\beta_n x) + \cos(\beta_n x),$$

where

$$\gamma_n = -\frac{e^{\beta_n} - \sin\beta_n + \cos\beta_n}{e^{\beta_n} + \sin\beta_n + \cos\beta_n} < 0.$$ (3.14)

By replacing x with x/l_i in the problem (3.10) and (3.11), we obtain the representation for λ_{in} and $u_{in}(x)$ in terms of β_n and $\phi_n(x/l_i)$:

$$\lambda_{in} = (\beta_n/l_i)^4, \quad u_{in}(x) = k_{in}\phi_n(x/l_i),$$ (3.15)

where the constants k_{in} are defined through the normalization of $u_{in}(x)$ in $L^2(0, l_i)$:

$$\|u_{in}\|^2_{L^2(0,l_i)} = \int_0^{l_i} u_{in}^2(x)dx = 1.$$

Hence,

$$k_{in} = \pm l_i^{-1/2} \left(\|\phi_n\|_{L^2(0,1)} \right)^{-1}. \tag{3.16}$$

By substituting (formally) expansion (3.9) into system of PDEs (3.8) and exploiting the orthogonality of $\{u_{in}(\cdot)\}$, we get the following infinite system of ODEs:

$$\ddot{\theta} = v,$$

$$\rho_i \ddot{q}_{in} = -\rho_i c_i^2 \lambda_{in} q_{in} + \rho_i \dot{\theta}^2 q_{in} - J_{in} v, \quad i = 1, 2, \ldots, k, \ n \in \mathbb{N}, \tag{3.17}$$

where

$$J_{in} = \int_0^{l_i} (x+d)u_{in}(x)\rho_i dx = \rho_i l_i^2 k_{in} \int_0^1 s\phi_n(s)ds + \rho_i l_i dk_{in} \int_0^1 \phi_n(s)ds. \tag{3.18}$$

Lemma 4.6 of [25, p. 176] implies that

$$\int_0^1 s\phi_n(s)ds = -2\beta_n^{-2} < 0, \quad \int_0^1 \phi_n(s)ds = 2\gamma_n \beta_n^{-1} < 0.$$

Therefore,

$$J_{in} = 2\rho_i l_i k_{in} \beta_n^{-1} \left(d\gamma_n - l_i \beta_n^{-1} \right) = O(n^{-1}) \neq 0. \tag{3.19}$$

By using expansion (3.9) for the kinetic (T) and potential (U) energy in formulas (3.2) and (3.3), we get:

$$2T = J\dot{\theta}^2 + \sum_{n=1}^{\infty} \sum_{i=1}^{k} \left(\rho_i (\dot{q}_{in}^2 + \dot{\theta}^2 q_{in}^2) + 2J_{in}\dot{\theta}\dot{q}_{in} \right),$$

$$2U = \sum_{n=1}^{\infty} \sum_{i=1}^{k} E_i I_i \lambda_{in} q_{in}^2,$$

where

$$J = J_0 + \sum_{i=1}^{k} \int_0^{l_i} (x+d)^2 \rho_i \, dx.$$

Note that, due to the definition of J_{in} in formula (3.18), the Parseval equality yields

$$\sum_{i=1}^{k} \int_0^{l_i} (x+d)^2 \, dx = \sum_{i=1}^{k} \sum_{n=1}^{\infty} \frac{J_{in}^2}{\rho_i^2}.$$

Hence,

$$J - \sum_{i=1}^{k} \sum_{n=1}^{\infty} \frac{J_{in}^2}{\rho_i} = J_0 > 0. \tag{3.20}$$

To define the state space for system (3.17), we assume that the total energy $E = T + U$ is finite on each solution of (3.17), i.e.

$$\sum_{i=1}^{k} \sum_{n=1}^{\infty} \lambda_{in}(q_{in})^2 < \infty, \quad \sum_{i=1}^{k} \sum_{n=1}^{\infty} \dot{q}_{in}^2 < \infty.$$

To satisfy these inequalities, we introduce new variables

$$\xi_{in} = c_i \sqrt{\lambda_{in}} q_{in}, \quad \eta_{in} = \dot{q}_{in}, \quad \omega = \dot{\theta},$$

and suppose that

$$\sum_{i=1}^{k} \sum_{n=1}^{\infty} (\xi_{in}^2 + \eta_{in}^2) < \infty. \tag{3.21}$$

With these variables, the system of differential equations (3.17) takes the form

$$\begin{aligned}
\dot{\theta} &= \omega, \quad \dot{\omega} = v, \\
\dot{\xi}_{in} &= c_i \sqrt{\lambda_{in}} \eta_{in}, \\
\dot{\eta}_{in} &= -c_i \sqrt{\lambda_{in}} \xi_{in} + \frac{\omega^2 \xi_{in}}{c_i \sqrt{\lambda_{in}}} - \frac{J_{in}}{\rho_i} v, \quad i = 1, 2, \ldots, k, \ n = 1, 2, \ldots
\end{aligned} \tag{3.22}$$

To write system (3.22) in the operator form, we introduce the infinite column vector

$$x = (\theta, \omega, \xi_{11}, \eta_{11}, \ldots, \xi_{k1}, \eta_{k1}, \xi_{12}, \eta_{12}, \ldots)^T.$$

The linear space of all columns that satisfy inequality (3.21) is denoted by ℓ^2. For $x \in \ell^2$ and

$$\tilde{x} = (\tilde{\theta}, \tilde{\omega}, \tilde{\xi}_{11}, \tilde{\eta}_{11}, \ldots, \tilde{\xi}_{k1}, \tilde{\eta}_{k1}, \tilde{\xi}_{12}, \tilde{\eta}_{12}, \ldots)^T \in \ell^2,$$

the inner product is defined as

$$\langle x, \tilde{x} \rangle = \theta \tilde{\theta} + \omega \tilde{\omega} + \sum_{i=1}^{k} \sum_{n=1}^{\infty} (\xi_{in} \tilde{\xi}_{in} + \eta_{in} \tilde{\eta}_{in}).$$

It is well-known that the linear space ℓ^2 with the inner product $\langle \cdot, \cdot \rangle$ is a Hilbert space.

Nonlinear system (3.22) is written in the operator form as follows:

$$\dot{x} = Ax + R(x)x + bv, \quad x \in \ell^2, \ v \in \mathbb{R}, \tag{3.23}$$

where $x = (\theta, \omega, \xi_{11}, \eta_{11}, \ldots, \xi_{k1}, \eta_{k1}, \xi_{12}, \eta_{12}, \ldots)^T \in \ell^2$ is the state vector and $v \in \mathbb{R}^1$ is the control. Here the unbounded linear operator $A : D(A) \to \ell^2$ is given by its block-diagonal matrix

$$A = \mathrm{diag}(A_0, A_{11}, \ldots, A_{k1}, A_{12}, \ldots, A_{k2}, \ldots),$$

$$A_0 = \begin{pmatrix} 0 & 1 \\ 0 & 0 \end{pmatrix}, \quad A_{in} = \begin{pmatrix} 0 & c_i \sqrt{\lambda_{in}} \\ -c_i \sqrt{\lambda_{in}} & 0 \end{pmatrix}, \quad (i = \overline{1, k}, \ n \in \mathbb{N}).$$

The infinite matrix of the operator $R(x)$ and the b column are also of the block form:

$$R(x) = \mathrm{diag}(O_{2 \times 2}, R_{11}(x), \ldots, R_{k1}(x), R_{12}(x), \ldots, R_{k2}, \ldots),$$

$$R_{in}(x) = \begin{pmatrix} 0 & 0 \\ \frac{\omega^2}{c_i \sqrt{\lambda_{in}}} & 0 \end{pmatrix},$$

$$b = (0, 1, \ 0, -\frac{J_{11}}{\rho_1}, \ 0, -\frac{J_{21}}{\rho_2}, \ \ldots, \ 0, -\frac{J_{k1}}{\rho_k}, \ 0, -\frac{J_{12}}{\rho_1}, \ \ldots)^T.$$

In formula (3.23), the standard multiplication of a matrix and a column is used. Since J_{in}/ρ_i are the Fourier coefficients of

$$\phi(x) = x + d \in L^2(0, l_i)$$

with respect to the basis $\{u_{in}(x)\}_{n=1}^{\infty}$, then

$$\sum_{n=1}^{\infty} J_{in}^2 < \infty$$

for all $1 \le i \le k$, so that $b \in \ell^2$.

The domain of definition $D(A) \subset \ell^2$ of the operator A consists of all $x \in \ell^2$ such that $Ax \in \ell^2$, i.e.

$$D(A) = \{x \in \ell^2 : \sum_{i=1}^{k} \sum_{n=1}^{\infty} c_i^2 \lambda_{in} (\eta_{in}^2 + \xi_{in}^2) < \infty\}.$$

In particular, $D(A)$ contains all vectors x having a finite number of non-zero coordinates, hence, $D(A)$ is dense in ℓ^2.

3.2 Stabilization by a State Feedback Law

In this section, we construct a feedback control which ensures strong asymptotic stability of the equilibrium point $x = 0$ for nonlinear system (3.23). Note that the operator A is not dissipative in ℓ^2 equipped with the standard norm, which means that the Lumer–Phillips theorem is not directly applicable in our case. Consequently, a direct application of known results on the stabilization of linear contractive [26] and spectral [27, 28] systems is not efficient.

To study nonlinear system (3.23), consider a Lyapunov functional of the total energy with some additional terms:

$$2V(x) = c_0 \theta^2 + J\omega^2 + \sum_{i=1}^{k} \sum_{n=1}^{\infty} \left(\rho_i \xi_{in}^2 + \rho_i \eta_{in}^2 + 2 J_{in} \omega \eta_{in} \right), \qquad (3.24)$$

where c_0 is an arbitrary positive constant. Let us write the Cauchy–Schwartz inequality in ℓ^2:

$$\left| \sum_{i=1}^{k} \sum_{n=1}^{\infty} J_{in} \eta_{in} \right| \le \left(\sum_{i,n} \frac{J_{in}^2}{\rho_i} \right)^{1/2} \left(\sum_{i,n} \eta_{in}^2 \rho_i \right)^{1/2}.$$

From this inequality and formula (3.24) it follows that

$$G\left(-|\omega|, \left(\sum_{i,n} \rho_i \eta_{in}^2 \right)^{1/2} \right) \le 2V(x) - c_0 \theta^2 - \sum_{i,n} \rho_i \xi_{in}^2$$

$$\le G\left(|\omega|, \left(\sum_{i,n} \rho_i \eta_{in}^2 \right)^{1/2} \right), \qquad (3.25)$$

where the quadratic form G is defined as

$$G(a, b) = Ja^2 + 2 \left(\sum_{i,n} \frac{J_{in}^2}{\rho_i} \right)^{1/2} ab + b^2.$$

We compute the principal minors of the quadratic form G:

$$\Delta_1 = J > 0, \quad \Delta_2 = J - \sum_{i,n} \frac{J_{in}^2}{\rho_i} = J_0 > 0.$$

We have used representation (3.20) here. According to Sylvester's criterion, G is a positive-definite quadratic form,

$$\alpha_1 (a^2 + b^2) \le G(a, b) \le \alpha_2 (a^2 + b^2), \tag{3.26}$$

with some positive constants $\alpha_1 > 0$ and $\alpha_2 > 0$.

Then representations (3.24) and (3.25) imply

$$\min\{c_0, \alpha_1, \rho_1, \ldots, \rho_k\} \|x\|^2 \le 2V(x) \le \max\{c_0, \alpha_2, \rho_1, \ldots, \rho_k\} \|x\|^2 \tag{3.27}$$

for all $x \in \ell^2$. Thus $V(x)$ is a positive-definite quadratic form.

By using the well-known rule to compute the differential of a functional in a Hilbert space, we write the time-derivative of V along the trajectories of system (3.23):

$$\dot{V} = \langle \nabla_x V, \dot{x} \rangle = \left(c_0 \theta + \sum_{i=1}^{k} \sum_{n=1}^{\infty} \frac{\xi_{in} \left(J_{in}(\omega^2 - c_i^2 \lambda_{in}) + \rho_i \omega \eta_{in} \right)}{c_i \sqrt{\lambda_{in}}} \right) \omega \tag{3.28}$$

$$+ \left(J - \sum_{i,n} \frac{J_{in}^2}{\rho_i} \right) \omega v.$$

Since $\dot{V}$ is divisible by ω, it is natural to choose a stabilizing feedback control $v = v(x)$ from the condition $\dot{V} = -h\omega^2 + o(\|x\|^2)$, where h is a positive constant. This condition yields the following linear functional:

$$v(x) = -\frac{1}{J_0} \left(c_0 \theta + h\omega - \sum_{i=1}^{k} \sum_{n=1}^{\infty} c_i \sqrt{\lambda_{in}} J_{in} \xi_{in} \right). \tag{3.29}$$

Then the time-derivative of V along the trajectories of (3.23) with $v = v(x)$ can be written as

$$\dot{V} = -h\omega^2 \left(1 - \frac{1}{h} \sum_{i=1}^{k} \sum_{n=1}^{\infty} \frac{\xi_{in}(J_{in}\omega + \rho_i \eta_{in})}{c_i \sqrt{\lambda_{in}}} \right). \tag{3.30}$$

Thus, the condition $\dot{V} \le 0$ is satisfied if

$$\sum_{i=1}^{k}\sum_{n=1}^{\infty}\frac{\xi_{in}(J_{in}\omega+\rho_i\eta_{in})}{c_i\sqrt{\lambda_{in}}} < h. \tag{3.31}$$

Using the formula (3.31) and evaluating (3.27), we conclude that the inequality $\dot{V}(x) \le 0$ holds for

$$x \in D(A) \cap X,$$

where

$$X = \left\{ x \in \ell^2 \mid V(x) \le \varkappa \right\} \tag{3.32}$$

with some constant $\varkappa > 0$.

In order to present stabilizability conditions for system (3.23), we need an auxiliary lemma.

Lemma 3.1 *Let*

$$c_i^2 \lambda_{in} \neq c_j^2 \lambda_{jm} \quad \textit{for all } i, j \in \{1, 2, \dots, k\}, \ n, m \in \mathbb{N}, \ (i, n) \neq (j, m). \tag{3.33}$$

Then for each $\tau > 0$ the system of functions

$$\{1, t, \sin(c_i\sqrt{\lambda_{in}}t), \cos(c_i\sqrt{\lambda_{in}}t) : t \in [0, \tau], \ i = \overline{1, k}, \ n = 1, 2, \dots\} \tag{3.34}$$

is linearly independent on $[0, \tau]$.

Proof We use Theorem 1.2.17 from [12] which can be formulated as follows: if

$$\limsup_{y\to\infty}\limsup_{z\to\infty}\frac{n[y, y+z)}{z} < \frac{\tau}{2\pi}, \tag{3.35}$$

then system (3.34) is minimal in $L^2(0, \tau)$. Here $n[a, b)$ stands for the cardinality of the set $[a, b) \cap \Lambda$,

$$\Lambda = \{c_i\sqrt{\lambda_{in}} : i = \overline{1, k}, \ n \in \mathbb{N}\}.$$

To prove (3.35), we note that $\lambda_{in} = (\beta_n/l_i)^4$, where β_n satisfies the transcendent equation of form (3.13), i.e.

$$\cos(\lambda_{in}^{1/4}l_i)\operatorname{ch}(\lambda_{in}^{1/4}l_i) = -1.$$

This implies the following asymptotic representation

$$\lambda_{in} = \left(\frac{\pi}{2l_i}\right)^4 (2n - 1)^4 + \Delta_{in}, \quad n = 1, 2, \dots, \tag{3.36}$$

where $\Delta_{in} \to 0$ as $n \to \infty$. By substituting (3.36) into (3.35), we get

$$\limsup_{y \to \infty} \limsup_{z \to \infty} \frac{n[y, y + z)}{z} = 0.$$

Now the assertion of lemma follows from Theorem 1.2.17 of [12].

We prove the following result.

Theorem 3.1 *Let the assumption* (3.33) *be satisfied and let* c_0, h *be positive constants. Then the feedback control* $v = v(x)$, *given by* (3.29), *ensures strong asymptotic stability of the trivial solution of system* (3.23), *that is, for any* $\varepsilon > 0$, *there exists a* $\delta = \delta(\varepsilon) > 0$ *such that each solution* $x(t)$ *of* (3.23) *with* $v = v(x)$, *satisfying the initial condition* $\|x(0)\| < \delta(\varepsilon)$, *has the following properties:*

1.
$$\|x(t)\| < \varepsilon, \quad \forall t \geq 0; \tag{3.37}$$

2.
$$\lim_{t \to +\infty} \|x(t)\| = 0. \tag{3.38}$$

Proof Consider a continuous functional $\mu(x) = \|x\|$ in ℓ^2. In order to prove the assertion of this Theorem, it suffices to verify conditions of Theorem 2.4 for the functional $V(x)$ of form (3.24) as the properties of strong asymptotic stability (3.37) and (3.38) are equivalent to asymptotic stability of the equilibrium $x = 0$ with respect to μ.

Condition (i) of Theorem 2.4 follows from inequality (3.27), and condition (ii) holds due to the construction of the set X in (3.32).

Let us represent functional (3.24) as $2V(x) = \langle Qx, x \rangle$, where Q is a self-adjoint operator. As $2V(x) = \langle Qx, x \rangle$ is positive definite then Q is positive, so it admits the following representation (cf. [29]):

$$Q = U^*U,$$

where the linear operator $U : l^2 \to l^2$ is bounded and $0 \notin \sigma(U)$. We define the following bilinear form in ℓ^2:

$$\langle x, \bar{x} \rangle_U = \langle Ux, U\bar{x} \rangle.$$

Owing to estimate (3.27), the standard norm

$$\|x\| = \left(\theta^2 + \omega^2 + \sum_{i=1}^{k} \sum_{n=1}^{\infty} (\eta_{in}^2 + \xi_{in}^2) \right)^{1/2}$$

and

$$\|x\|_U = \sqrt{\langle x, x \rangle_U}.$$

are equivalent in ℓ^2. Let us write the linear part of system (3.23) with $v = v(x)$ in the form

$$\dot{x} = -\tilde{A}x, \quad x(0) = x_0 \in \ell^2, \qquad (3.39)$$

where the domain of definition $D(\tilde{A})$ is dense in ℓ^2, $\tilde{A}(0) = 0$. Let us compute the time-derivative of the functional $V(x) = \frac{1}{2}\langle Qx, x \rangle$ along the trajectories of linear differential equation (3.39):

$$\dot{V}(x) = \langle Qx, -\tilde{A}x \rangle = -h\omega^2 \le 0,$$

if $x \in D(\tilde{A})$. Besides that,

$$\langle Qx, -\tilde{A}x \rangle = \langle Ux, -U\tilde{A}x \rangle = -\langle x, \tilde{A}x \rangle_U.$$

It means that the linear operator $\tilde{A} : D(\tilde{A}) \to \ell^2$ is accretive in ℓ^2 with the inner product $\langle \cdot, \cdot \rangle_U$, i.e. the operator $-\tilde{A} : D(\tilde{A}) \to \ell^2$ is dissipative. To check condition (iii) of Theorem 2.4, let us establish the precompactness of the trajectories of linear differential equation (3.39) by the Dafermos–Slemrod Theorem (Theorem 1.6). For this purpose we show that the operator $(\lambda\tilde{A} + I)^{-1} : \ell^2 \to \ell^2$ is compact for some $\lambda > 0$.

Consider the equation $Ix - \lambda Ax - \lambda bv = \bar{x}$ with respect to x, where $\lambda = \text{const}$,

$$x = (\theta, \omega, \xi_{11}, \eta_{11}, \dots, \xi_{k1}, \eta_{k1}, \xi_{12}, \eta_{12}, \dots)^T \in \ell^2,$$
$$\bar{x} = (\bar{\theta}, \bar{\omega}, \bar{\xi}_{11}, \bar{\eta}_{11}, \dots, \bar{\xi}_{k1}, \bar{\eta}_{k1}, \bar{\xi}_{12}, \bar{\eta}_{12}, \dots)^T \in \ell^2.$$

In the coordinate form, this equation becomes

$$(I_{2\times2} - \lambda A_0)\begin{pmatrix} \theta \\ \omega \end{pmatrix} = \begin{pmatrix} \bar{\theta} \\ \bar{\omega} + \lambda v \end{pmatrix},$$

$$(I_{2\times2} - \lambda A_{in})\begin{pmatrix} \xi_{in} \\ \eta_{in} \end{pmatrix} = \begin{pmatrix} \bar{\xi}_{in} \\ \bar{\eta}_{in} - \lambda \frac{J_{in}}{\rho_i}v \end{pmatrix}, \quad i = 1, 2, \dots, k, \ n = 1, 2, \dots,$$

where A_0 and A_{in} are 2×2-matrices. By using the inverse matrix method to solve these equations, we obtain

$$\omega = \bar{\omega} + \lambda v, \quad \theta = \bar{\theta} + \lambda\bar{\omega} + \lambda^2 v, \qquad (3.40)$$

$$\begin{pmatrix} \xi_{in} \\ \eta_{in} \end{pmatrix} = \frac{1}{1 + \lambda^2 c_i^2 \lambda_{in}}\begin{pmatrix} 1 & \lambda_i c_i \sqrt{\lambda_{in}} \\ -\lambda c_i \sqrt{\lambda_{in}} & 1 \end{pmatrix} \times \begin{pmatrix} \bar{\xi}_{in} \\ \bar{\eta}_{in} - \frac{\lambda J_{in}}{\rho_i}v \end{pmatrix}. \qquad (3.41)$$

Then the substitution of expressions (3.40) and (3.41) into v given by (3.29) yields

$$v = - \left(J_0 + \lambda h + \lambda^2 c_0 + \lambda^2 \sum_{i=1}^{k} \sum_{n=1}^{\infty} \frac{c_i^2 \lambda_{in} J_{in}^2}{\rho_i (1 + \lambda^2 c_i^2 \lambda_{in})} \right)^{-1}$$

$$\times \left(c_0 \bar{\theta} + (\lambda c_0 + h) \bar{\omega} - \sum_{i=1}^{k} \sum_{n=1}^{\infty} \frac{c_i J_{in} \sqrt{\lambda_{in}} (\bar{\xi}_{in} + \lambda c_i \sqrt{\lambda_{in}} \bar{\eta}_{in})}{1 + \lambda^2 c_i^2 \lambda_{in}} \right). \quad (3.42)$$

Formula (3.42) defines a linear functional $v(\bar{x})$ in ℓ^2. For an arbitrary $\lambda > 0$, the functional $v(\bar{x})$ is bounded as its coefficients belong to ℓ^2:

$$0 < J_0 + \lambda h + \lambda^2 c_0 + \lambda^2 \sum_{i=1}^{k} \sum_{n=1}^{\infty} \frac{c_i^2 \lambda_{in} J_{in}^2}{\rho_i (1 + \lambda^2 c_i^2 \lambda_{in})} = \text{const} < \infty,$$

$$\sum_{i=1}^{k} \sum_{n=1}^{\infty} \frac{J_{in}^2 \lambda_{in}}{(1 + \lambda^2 c_i^2 \lambda_{in})^2} < \infty, \qquad \sum_{i=1}^{k} \sum_{n=1}^{\infty} \frac{J_{in}^2 \lambda_{in}^2}{(1 + \lambda^2 c_i^2 \lambda_{in})^2} < \infty.$$

The above series converge due to estimates (3.12) and (3.19). Therefore, for any $\lambda > 0$, there is a positive $M_1(\lambda)$ such that $|v(\bar{x})| \leq M_1(\lambda) \|\bar{x}\|$ in formula (3.42) for all $\bar{x} \in \ell^2$.

Formulas (3.40)–(3.42) define $x = (\lambda \tilde{A} + I)^{-1} \bar{x}$ for all $\bar{x} \in \ell^2$ if $\lambda > 0$. This implies

$$\|(\lambda \tilde{A} + I)^{-1} \bar{x}\|^2 \leq (\bar{\omega} + \lambda v(\bar{x}))^2 + (\bar{\theta} + \lambda \bar{\omega} + \lambda^2 v(\bar{x}))^2$$

$$+ 2 \left(\sum_{i,n} \frac{1}{1 + \lambda^2 c_i^2 \lambda_{in}} \right) \sum_{i,n} \left(\bar{\xi}_{in}^2 + (\bar{\eta}_{in} - \lambda J_{in} v(\bar{x})/\rho_i)^2 \right)$$

$$\leq M_2(\lambda) \|\bar{x}\|^2$$

for some constant $M_2(\lambda)$ because

$$\sum_{i=1}^{k} \sum_{n=1}^{\infty} \frac{1}{1 + \lambda^2 c_i^2 \lambda_{in}} < \infty, \quad (3.43)$$

and the functional $v(\bar{x})$ is bounded.

Hence, the following bounded linear operator

$$(\lambda \tilde{A} + I)^{-1} : \ell^2 \to \ell^2$$

is defined for each $\lambda > 0$. To prove that the above operator is compact, we consider $\Pi_N : \ell^2 \to \ell^2$ that projects the elements $x \in \ell^2$ onto the subspace with $\theta = \omega = \xi_{in} = \eta_{in} = 0$ for $n < N$. In the coordinate form, the projection operator Π_N may be written as follows:

$$\Pi_N x = (0, 0, \ldots, 0, \, \xi_{1N}, \eta_{1N}, \ldots, \xi_{kN}, \eta_{kN}, \, \xi_{1,N+1}, \eta_{1,N+1}, \ldots)^T.$$

Consider a family of bounded linear operators in ℓ^2:

$$U_N = (I - \Pi_N)(\lambda \tilde{A} + I)^{-1}.$$

Each operator U_N is compact as its image is of finite dimension. We will show that $(\lambda \tilde{A} + I)^{-1}$ is the limit of the following operators:

$$\lim_{N \to \infty} \|(\lambda \tilde{A} + I)^{-1} - U_N\| = \lim_{N \to \infty} \|\Pi_N(\lambda \tilde{A} + I)^{-1}\| = 0. \tag{3.44}$$

We have:

$$
\begin{aligned}
\|\Pi_N(\lambda \tilde{A} + I)^{-1}\bar{x}\|^2 &\leq 2 \left(\sum_{i=1}^{k} \sum_{n=N}^{\infty} \frac{1}{1 + \lambda^2 c_i^2 \lambda_{in}} \right) \sum_{i=1}^{k} \sum_{n=1}^{\infty} \left(\bar{\xi}_{in}^2 + (\bar{\eta}_{in} - \lambda J_{in} v(\bar{x})/\rho_i)^2 \right) \\
&\leq 2 \left(\sum_{i=1}^{k} \sum_{n=N}^{\infty} \frac{1}{1 + \lambda^2 c_i^2 \lambda_{in}} \right) \\
&\quad \times \left(1 + 2\lambda M_1(\lambda) \left(\sum_{i=1}^{k} \sum_{n=1}^{\infty} \frac{J_{in}^2}{\rho_i^2} \right)^{1/2} + \lambda^2 M_1(\lambda)^2 \sum_{i=1}^{k} \sum_{n=1}^{\infty} \frac{J_{in}^2}{\rho_i^2} \right) \|\bar{x}\|^2.
\end{aligned}
\tag{3.45}
$$

The convergence of series (3.43) implies that

$$\sum_{n=N}^{\infty} \frac{1}{1 + \lambda^2 c_i^2 \lambda_{in}} \to 0 \qquad N \to \infty.$$

Hence, estimate (3.45) yields the property (3.44), so that the operator $(\lambda \tilde{A} + I)^{-1} : \ell^2 \to \ell^2$ is compact as the limit of finite-dimensional operators (cf. [30]).

In particular, the compactness of the linear operator $(\lambda \tilde{A} + I)^{-1} : \ell^2 \to \ell^2$ implies that the range $\mathscr{R}(\lambda \tilde{A} + I)$ of $\lambda \tilde{A} + I$ is ℓ^2, and the operator $-\tilde{A}$ is m-dissipative (in the sense of Definition 1.8). The Lumer–Phillips theorem (Theorem 1.3) implies that $-\tilde{A}$ is the infinitesimal generator of a C_0-semigroup $\{e^{-t\tilde{A}}\}_{t \geq 0}$ of contractions on ℓ^2 with the norm $\|\cdot\|_U$. As the operator $(\lambda \tilde{A} + I)^{-1}$ is compact, Theorem 1.6 implies that the trajectories of linear differential equation (3.39) are precompact in ℓ^2.

Now we represent the nonlinear closed-loop system (3.22) with control (3.29) as follows:

$$\dot{x}(t) = -\tilde{A}x(t) + \omega^2(t)Cx(t), \quad x(0) = x_0 \in \ell^2, \tag{3.46}$$

where

$$Cx = \left(0, 0, 0, \frac{\xi_{11}}{c_1\sqrt{\lambda_{11}}}, \dots, 0, \frac{\xi_{k1}}{c_k\sqrt{\lambda_{k1}}}, 0, \frac{\xi_{12}}{c_1\sqrt{\lambda_{12}}}, \dots, 0, \frac{\xi_{k2}}{c_k\sqrt{\lambda_{k2}}}, \dots\right)^T.$$

(3.47)

Note that the right-hand side of equation (3.46) is a Lipschitz perturbation of the operator $-\tilde{A}$, therefore, for any $x_0 \in \ell^2$, there is a unique mild solution $x(t)$ of problem (3.46) on $t \in [0, t_{max})$ by Theorem 1.4 of [31, p. 185]. The set X in (3.32) is (forward) invariant by the construction. As X is bounded then the initial condition $x_0 \in X$ implies that $x(t)$ is bounded, hence $t_{max} = +\infty$ by Theorem 1.4 of [31, p. 185]. Therefore, nonlinear differential equation (3.46) generates a continuous semigroup of operators $\{S(t)\}_{t \geq 0}$ on X.

Let us apply Theorem 2.5 to prove that the trajectories of equation (3.46) are precompact. We assume

$$f(t) = \omega^2(t), \quad R(x, t) = Cx(t)$$

in conditions of Theorem 2.5. As the time-derivative of V along the trajectories of (3.46) admits the estimate

$$\dot{V}(x(t)) \leq -h^*\omega^2(t), \quad h^* = const > 0,$$

(3.48)

for $\|x\| \leq \mu$, then

$$\int_0^{+\infty} \omega^2(t)dt < \infty.$$

(3.49)

Thus, $f \in L^1[0, +\infty)$ provided that $x_0 \in X$. The representation (3.36) implies that

$$\sum_{n=1}^{\infty} \frac{1}{\lambda_{in}} < \infty,$$

for each $i = 1, 2, \dots, k$, i.e. the linear operator $C : \ell^2 \to \ell^2$ is compact as its matrix norm is finite. By taking into account the boundedness of $x(t) \in X$, we conclude that $R(x, t)$ takes values in a compact subset if $x_0 \in X$. It has been already proved that the trajectories $\{e^{-t\tilde{A}}x_0\}_{t \geq 0}$ of the linearized equation are precompact, so the trajectories of nonlinear equation (3.46) are precompact by Theorem 2.5.

It remains to verify condition (v) of Theorem 2.4. Let us show that the set

$$M = \{x \in \ell^2 \mid \dot{V}(x) = 0\}$$

does not contain any nontrivial trajectory of system (3.22) with $v = v(x)$, $t \geq 0$. Since $\dot{V}(x)$ satisfies inequality (3.48) then each trajectory from M should satisfy the following relations

$$\omega(t) = 0, \quad \theta(t) = \theta(0) + \omega(0)t, \quad v(x(t)) = \dot{\omega}(t) = 0,$$

$$\theta(0) = \sum_{i=1}^{k} \sum_{n=1}^{\infty} c_i \sqrt{\lambda_{in}} J_{in} \xi_{in}(t),$$

$$\xi_{in}(t) = \xi_{in}(0) \cos(c_i \sqrt{\lambda_{in}} t) + \eta_{in}(0) \sin(c_i \sqrt{\lambda_{in}} t),$$

$$\eta_{in}(t) = -\xi_{in}(0) \sin(c_i \sqrt{\lambda_{in}} t) + \eta_{in}(0) \cos(c_i \sqrt{\lambda_{in}} t),$$

In particular, the above expressions imply

$$\theta(0) \equiv \sum_{i=1}^{k} \sum_{n=1}^{\infty} c_i \sqrt{\lambda_{in}} J_{in} \left(\xi_{in}(0) \cos(c_i \sqrt{\lambda_{in}} t) + \eta_{in}(0) \sin(c_i \sqrt{\lambda_{in}} t) \right) \quad (3.50)$$

for all $t \geq 0$. From Lemma 3.1 and conditions $J_{in} \neq 0$, $\lambda_{in} > 0$ in (3.19) it follows that the identity (3.50) holds on M iff $x(0) = 0$ under the conditions of Assumption (3.33). Therefore, the set M does not contain any nontrivial trajectory, so the equilibrium $x = 0$ of system (3.22) with $v = v(x)$ is strongly asymptotically stable by Theorem 2.4.

3.3 Partial Stabilization in the Case of Multiple Frequencies

In this section, we investigate the stabilization problem for system (3.23) if the assumption (3.33) is violated. To simplify this study, we assume that all beams have the same mechanical properties, i.e.

$$l_i = l, \quad \rho_i = \rho, \quad c_i = c$$

for all $i = 1, 2, \ldots, k$. Then $\lambda_{1n} = \cdots = \lambda_{kn} = \lambda_n$ in the representation (3.15). In this case, we show that the linear approximation of (3.23) is not stabilizable if $k \geq 2$. Indeed, consider a nonsingular linear transformation $\Phi : \ell^2 \to \ell^2$ as follows:

$$\Phi(x) = (\theta, \omega, \ldots, Q_{in}, P_{in}, \ldots)^T,$$

where

$$Q_{kn} = \frac{1}{k} \sum_{j=1}^{k} \xi_{jn}, \quad P_{kn} = \frac{1}{k} \sum_{j=1}^{k} \eta_{jn},$$

$$Q_{in} = \xi_{in} - Q_{kn}, \quad P_{in} = \eta_{in} - P_{kn}, \quad i = \overline{1, k-1}.$$

By applying Φ to the linear party of system (3.23), we obtain the following differential equations with respect to Q_{in} and P_{in}:

$$\dot{Q}_{in} = c\sqrt{\lambda_n}\,P_{in}, \quad \dot{P}_{in} = -c\sqrt{\lambda_n}\,Q_{in}, \quad (i = \overline{1, k-1}, \; n \geq 1). \tag{3.51}$$

We see that, for each solution of Eq. (3.51), the following identity holds:

$$\sum_{i=1}^{k-1} \sum_{n=1}^{\infty} \left(P_{in}(t)^2 + Q_{in}(t)^2 \right) = \text{const.} \tag{3.52}$$

The above consideration implies that the trivial solution of subsystem (3.51) cannot be made asymptotically stable if $k \geq 2$. We show that the partial asymptotic stabilization of system (3.23) is possible in this case.

To consider the partial stabilization problem, we define a bounded linear operator $\Pi : \ell^2 \to \ell^2$ as follows:

$$\Pi : x \mapsto (\theta, \omega, \sum_{i=1}^{k} \xi_{i1}, \sum_{i=1}^{k} \eta_{i1}, \ldots, \sum_{i=1}^{k} \xi_{in}, \sum_{i=1}^{k} \eta_{in}, \ldots)^T. \tag{3.53}$$

In $k = 1$ then Π is the identity operator in ℓ^2, otherwise the value of

$$\| \Pi \Phi^{-1}(\theta, \omega, \ldots, Q_{in}, P_{in}, \ldots,)^T \|$$

does not depend on Q_{in} and P_{in} for $i \leq k-1$. Hence the constant of the integral (3.52) does not affect the value $\| \Pi\, x(t) \|$ in the linear case.

Formula (3.53) shows that the operator Π projects the state space ℓ^2 onto the linear subspace parameterized with the rotation angle, the angular velocity of the body, and coordinates Q_{kn}, P_{kn}. Therefore, the vector Πx describes the state of an "averaged" mechanical system with a single beam, for which the generalized coordinates with index n are obtained by averaging the modal coordinates with the same index n for all k beams of the original system.

Consider a continuous functional $\mu(x) = \| \Pi x \|$ in ℓ^2. From a mechanical viewpoint, the stabilization of system (3.23) with respect to the functional μ leads to the stabilization of the attitude of the rigid body together with damping of the vibrations corresponding to averaged modal coordinates.

The basic result concerning the partial stabilization problem of a flexible system with resonances is as follows.

Theorem 3.2 *Let c_0 and h be arbitrary positive constants. Then the feedback control $v = v(x)$ of form (3.29) ensures asymptotic stability of the solution $x = 0$ of nonlinear system (3.23) with respect to the functional*

$$\mu(x) = \| \Pi x \|$$

in ℓ^2. Moreover, the solution $x = 0$ of the closed-loop system (3.23) and (3.29) is (non-asymptotically) stable in the sense of Lyapunov.

Proof The estimate (3.27) implies that the functional V satisfies condition (i) of Theorem 2.4 with $\mu(x) = \|\Pi x\|$. We see that $\dot{V}(x) \le 0$ for all $x \in D(A) \cap X$ if the set X is defined by formula (3.32). The precompatness of the trajectories of nonlinear system (3.23) with the feedback control (3.29) has been established in the proof of Theorem 3.1. Thus, it remains to verify conditions (iv) and (v) of Theorem 2.4.

The closed-loop system (3.23) with $v = v(x)$ on the set

$$Z = \overline{\{x \in D(A) \mid \dot{V}(x) = 0\}}$$

takes the following form:

$$\dot{x}(t) = Ax(t), \quad \omega(t) = 0, \quad v(x(t)) = 0.$$

By solving this linear equation, we get

$$\theta(t) = \theta_0,$$
$$\xi_{in}(t) = C_{in}^{(1)} \sin(\sqrt{\lambda_n}ct) + C_{in}^{(2)} \cos(\sqrt{\lambda_n}ct), \tag{3.54}$$
$$\eta_{in}(t)V = C_{in}^{(1)} \cos(\sqrt{\lambda_n}ct) - C_{in}^{(2)} \sin(\sqrt{\lambda_n}ct),$$

where the constants θ_0 and $C_{in}^{(j)}$ are defined by the initial condition. As the left-hand side of (3.29) vanishes on Z, the substitution of (3.54) into (3.29) yields:

$$\frac{c_0\theta_0}{c} = \sum_{n=1}^{\infty} \sum_{i=1}^{k} J_n\sqrt{\lambda_n} \left(C_{in}^{(1)} \sin(\sqrt{\lambda_n}ct) + C_{in}^{(2)} \cos(\sqrt{\lambda_n}ct) \right). \tag{3.55}$$

Let us assume that the closed-loop system admits a trajectory on Z for $t \ge 0$. Then relation (3.55) holds with some constants θ_0 and $C_{in}^{(j)}$. The system of functions

$$\{1, \sin\sqrt{\lambda_n}\tau, \cos\sqrt{\lambda_n}\tau \mid \tau \ge 0, n \in \mathbb{N}\}$$

is linearly independent on $\mathbb{R}^+$ by Theorem 1.2.17 of [12] because of estimate (3.12). Since $\lambda_n \ne 0$ and $J_n = J_{in} \ne 0$ in (3.19), then relation (3.55) holds only if

$$\theta_0 = \sum_{i=1}^{k} C_{in}^{(1)} = \sum_{i=1}^{k} C_{in}^{(2)} = 0, \quad \forall n \in \mathbb{N}. \tag{3.56}$$

This means that the maximal invariant subset of Z is contained in

$$Z_0 = \{x \in \ell^2 \mid \theta = \omega = \sum_{i=1}^{k} \xi_{in} = \sum_{i=1}^{k} \eta_{in} = 0, \forall n \in \mathbb{N}\}.$$

The property Ker $\Pi = Z_0$ implies that conditions (iv) and (v) of Theorem 2.4 hold. Thus, the solution $x = 0$ of the closed-loop system (3.23) with (3.29) is asymptotically stable with respect to the functional μ by Theorem 2.4.

3.4 Numerical Example: Stabilization of a Finite-Dimensional System

Let us fix a finite number N and consider a subsystem of nonlinear system (3.22) with modal coordinates of indices $n \le N$. If we also discard terms ξ_{in} and η_{in} with $n > N$ in the feedback control (3.29), we obtain a finite dimensional approximation of the closed-loop system.

Figures 3.2 and 3.3 show simulation results for the finite dimensional approximation of system (3.22) and (3.29) for $N = 3$ and $k = 2$ with control parameters $c_0 = 0.1$, $h = 0.1$, and the following values of mechanical parameters:

$$J_0 = 0.1\,\text{kg} \cdot \text{m}^2, \quad d = 0.1\,\text{m}, \quad l_i = 1\,\text{m}, \quad \rho_i = 0.0675\,\text{kg/m},$$
$$c_i^2 = 54.6\,\text{N} \cdot \text{m}^3/\text{kg}, \quad i \le 2.$$

The initial conditions are chosen as follows:

$$\theta(0) = \pi/2, \quad \omega(0) = 0,$$
$$\xi_{in}(0) = \eta_{in}(0) = 0 \ (i = \overline{1,2}, \ n = \overline{1,3}).$$

The above choice of mechanical parameters corresponds to the model with two aluminium beams of the square cross-section with the square side 1 cm.

We see that the controller (3.29) stabilizes the attitude of the rigid body with these values of parameters (Fig. 3.2). The displacement of the free end-point of a beam is shown in Fig. 3.3. As the beams and initial conditions are identical for $i = 1$ and $i = 2$, we present here only the graph for

$$w_1(l_1, t) = \sum_{n=1}^{N} u_{1n}(l_1)q_{1n}(t), \quad (N = 3).$$

Thus, Fig. 3.3 illustrates damped oscillations of the free end-point for each beam.

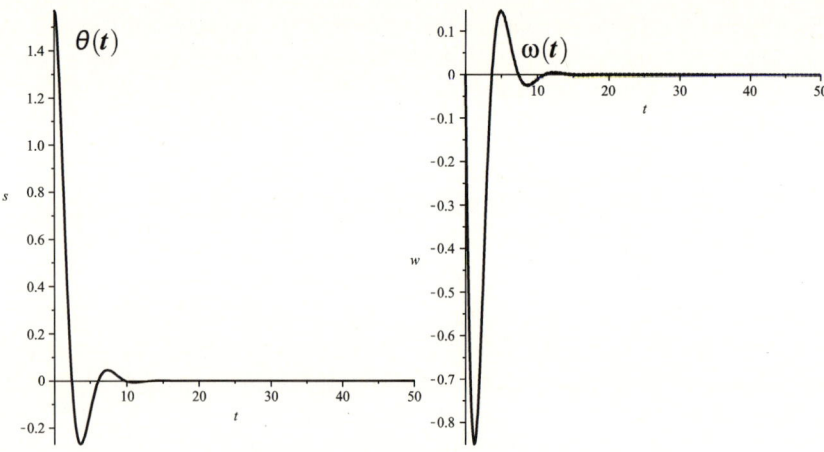

Fig. 3.2 Attitude and the angular velocity of the rigid body

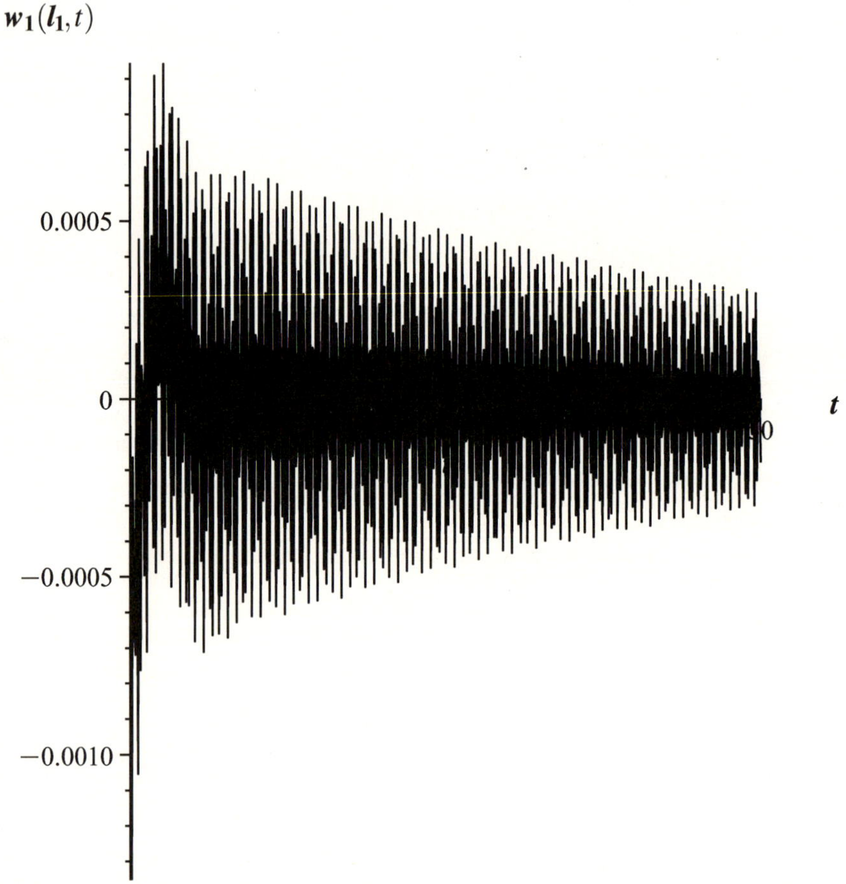

$w_1(l_1, t)$

Fig. 3.3 Time plot of the function $w_1(l_1, t)$

3.5 Stabilization of a Flexible-Link Manipulator Model

The goal of this section is to study the stabilization problem for a multi-link manipulator model consisting of rigid bodies and Euler–Bernoulli beams.

3.5.1 Description of the Model

Consider a mechanical system consisting of rigid bodies B_0, B_1, Σ, and a chain of flexible links $S_1, S_2, \ldots, S_n$ (Fig. 3.4). We assume that rigid bodies B_0 and B_1 are joined by a cylindrical hinge, and that B_1 carries n flexible beams $S_1, S_2, \ldots,$ S_n together with the body Σ. Let $OXYZ$ be a fixed Cartesian frame with the unit vectors (h_1, h_2, h_3), and let the body B_0 rotates around the vertical axis OZ. Denote by $0_1 x_1 y z_1$ and $0_1 x y z$ the Cartesian frames related to bodies B_0 and B_1, respectively. Note that $0_1 x_1 y z_1$ is obtained by the translation of the $OXYZ$ frame along the OX-axis by $-R$ and subsequent rotation around the OZ-axis by $-\varphi_T$. The frame $O_1 x y z$ is obtained by the rotation of $O_1 x_1 y z_1$ around the $O_1 y$-axis by angle $-\varphi_R$. Thus, the relation between fixed vectors (h_1, h_2, h_3) and unit vectors (e_1, e_2, e_3) of the frame $O_1 x y z$ is as follows:

$$
\begin{pmatrix} e_1 \\ e_2 \\ e_3 \end{pmatrix} = \begin{pmatrix} \cos\varphi_T \cos\varphi_R & -\sin\varphi_T \cos\varphi_R & \sin\varphi_R \\ \sin\varphi_T & \cos\varphi_T & 0 \\ -\cos\varphi_T \sin\varphi_R & \sin\varphi_T \sin\varphi_R & \cos\varphi_R \end{pmatrix} \cdot \begin{pmatrix} h_1 \\ h_2 \\ h_3 \end{pmatrix}. \tag{3.57}
$$

The angular velocity ω of the frame $O_1 x y z$ as an absolutely rigid body is

$$
\omega = -\dot{\varphi}_T \sin\varphi_R e_1 - \dot{\varphi}_R e_2 - \dot{\varphi}_T \cos\varphi_R e_3. \tag{3.58}
$$

The link S_1 is clamped at the point O_1 of the body B_1. Each link S_j, $j = 2, \ldots, n$ may perform the translational motion with respect to the previous link S_{j-1} (as a telescoping boom). The endpoint O_2 of the link S_n is clamped at the rigid body Σ (payload).

Let l be the distance from the point O_1 to the orthogonal projection of the point O_2 onto the $O_1 x$-axis, $l \in [l_{min}, l_{max}]$ (i.e. l is a generalized coordinate characterizing the length of the system of links). We assume that the thickness of each link S_j, $j = 1, 2, \ldots, n$ may be neglected in comparison with its length l_j, and the reference centerline of S_j coincides with the segment $[a_j(l), b_j(l)]$ on the $O_1 x$-axis, where

$$
a_j(l) = \begin{cases} 0, & n = 1, \\ \frac{j-1}{n-1}(l - l_{min}) + a_{j,min}, & n > 1, \end{cases} \qquad b_j(l) = a_j(l) + l_j, \tag{3.59}
$$

and $a_{j,min} = $ const is the distance from O_1 to the nearest point of the link S_j in its undeformed state as $l = l_{min}$. Formula (3.59) is obtained by assuming that the

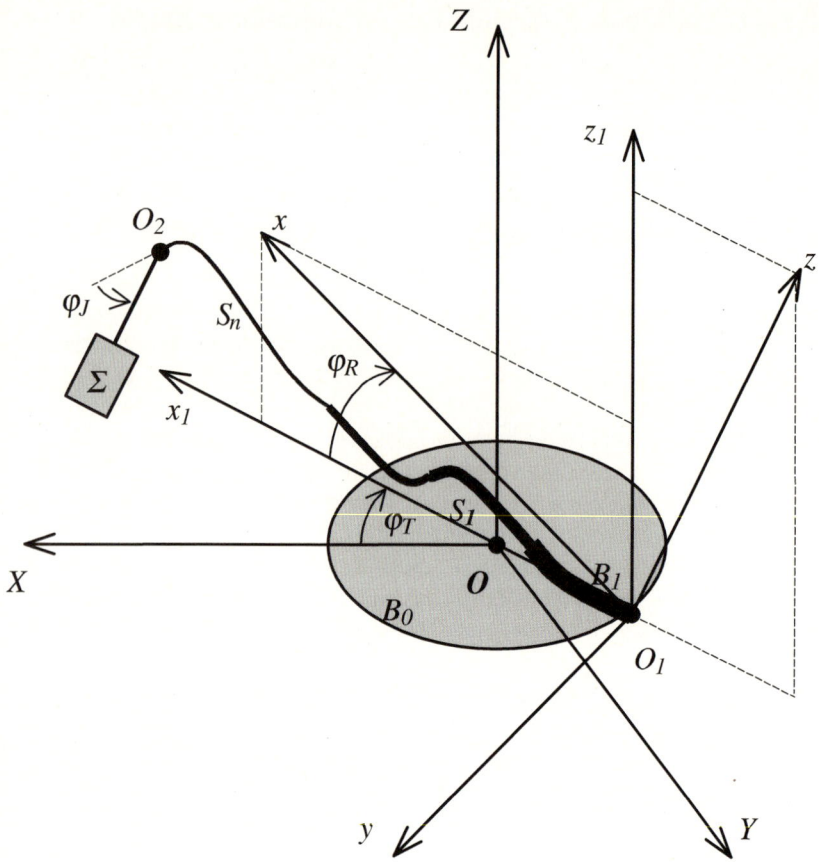

Fig. 3.4 Flexible-link manipulator model

longitudinal displacements of S_j with respect to S_{j-1} are equal for all $j = 2, \ldots, n$. In practice, such motion may be implemented by a rope and pulleys attached to the links.

With each point M of the link S_j we associate its Lagrangian coordinate $\xi_j \in [0, l_j]$ as the distance between M and the endpoint of S_j (which is nearest to the point O_1) in its undeformed state. Thus, under assumptions of the thin beams theory, the position of a point M at time t is described as follows:

$$O_1 M = (a_j(l) + \xi_j)e_1 + y_j(\xi_j, t)e_2 + z_j(\xi_j, t)e_3, \quad \xi_j \in [0, l_j], \ t \geq 0.$$

In addition, the following contact condition should be satisfied between the links:

$$y_j(x - a_j(l), t) = y_{j-1}(x - a_{j-1}(l), t), \qquad (3.60)$$

$$z_j(x - a_j(l), t) = z_{j-1}(x - a_{j-1}(l), t), \quad \forall x \in [b_{j-1}(l), a_j(l)], \ j = \overline{2, n}.$$

Let us compute the absolute velocity of a point $M \in S_j$ with the Lagrangian coordinate ξ_j:

$$v_M(\xi_j, t) = R\dot{\varphi}_T e_2 + \omega \times O_1 M + a_j(l(t))\dot{e}_1 + \dot{y}_j(\xi_j, t)e_2 + \dot{z}_j(\xi_j, t)e_3$$

$$= \left(\frac{j-1}{n-1} i + \dot{\varphi}_T y_j \cos \varphi_R - \dot{\varphi}_R z_j \right) e_1$$

$$+ \left(\dot{y}_j + \dot{\varphi}_T [R + z_j \sin \varphi_R - (a_j + \xi_j) \cos \varphi_R] \right) e_2$$

$$+ \left(\dot{z}_j + (a_j + \xi_j)\dot{\varphi}_R - y_j \dot{\varphi}_T \sin \varphi_R \right) e_3. \qquad (3.61)$$

Here, and in the sequel, all fractional expressions with $n-1$ in the denominator are assumed to be zero for $j = n = 1$. The kinetic energy of the link S_j, $j = \overline{1, n}$ is:

$$2T_j = \int_0^{l_j} \rho_j(\xi_j) \left\{ (y_j \dot{\varphi}_T \cos \varphi_R - z_j \dot{\varphi}_R)^2 + \left(\frac{j-1}{n-1} \right)^2 i^2 \right.$$

$$+ \frac{2(j-1)i}{n-1} (y_j \dot{\varphi}_T \cos \varphi_R - z_j \dot{\varphi}_R)$$

$$+ (\dot{y}_j + \dot{\varphi}_T [R + z_j \sin \varphi_R - (a_j(l) + \xi_j) \cos \varphi_R])^2$$

$$+ \left. (\dot{z}_j - y_j \dot{\varphi}_T \sin \varphi_R + (a_j(l) + \xi_j)\dot{\varphi}_R)^2 \right\} d\xi_j, \qquad (3.62)$$

where $\rho_j(\xi_j)$ is the mass per unit length (linear density) of the link S_j. Within the Euler–Bernoulli beam theory, the potential energy of S_j has the form

$$2\Pi_j = \int_0^{l_j} \left\{ E_j \mathscr{I}_{jz}(y_j'')^2 + E_j \mathscr{I}_{jy}(z_j'')^2 + 2g\rho_j(\xi_j)(z_j \cos \varphi_R + (a_j(l) + \xi_j) \sin \varphi_R) \right\} d\xi_j,$$

$$(3.63)$$

$j = \overline{1, n}$, where the prime denotes the differentiation with respect to ξ_j; E_j is Young's modulus; $\mathscr{I}_{jy}$ and $\mathscr{I}_{jz}$ are central moments of inertia of the cross-section area of the jth link with respect to axes parallel to O_1 and $O_1 z$, respectively; and g is the gravity constant. Suppose that E_j, $\mathscr{I}_{jy}$, and $\mathscr{I}_{jz}$ are positive C^2-functions of $\xi_j \in [0, l_j]$.

The kinetic and potential energy of B_0 and B_1 is the following

$$2T_0 = (I_0 + I_1 \sin^2 \varphi_R + I_3 \cos^2 \varphi_R + m_0(R - d \cos \varphi_R)^2)\dot{\varphi}_T^2 + (I_2 + m_0 d^2)\dot{\varphi}_R^2,$$

$$\Pi_0 = m_0 g d \sin \varphi_R. \qquad (3.64)$$

Here I_0 is the moment of inertia of B_0 with respect to the OZ-axis; I_1, I_2, and I_3 are central moments of inertia of B_1 with respect to the axes parallel to $O_1 x_1$, $O_1 y$, and

$O_1 z_1$, respectively; and m_0 is the mass of B_1. We assume that the coordinate axes of $O_1 x_1 y z_1$ are principal axes of inertia of B_1, and the center of mass of B_1 is on the $O x_1$-axis at the distance d from O_1.

In order to describe the motion of the payload Σ, we introduce a trihedron $O_2 v_1 v_2 v_3$ rigidly connected to the link S_n at the point O_2. In the undeformed state, the vectors (v_1, v_2, v_3) are collinear to (e_1, e_2, e_3). The body Σ rotates in a cylindrical hinge around $O_2 v_2$ by the angle φ_J. With the rigid body Σ, we associate a trihedron $O_2 \gamma_1 \gamma_2 \gamma_3$ obtained by the rotation of $O_2 v_1 v_2 v_3$ around $O_2 v_2$ by the angle φ_J. The basis $(\gamma_1, \gamma_2, \gamma_3)$ is defined by the following formulas:

$$\begin{pmatrix} \gamma_1 \\ \gamma_2 \\ \gamma_3 \end{pmatrix} = \begin{pmatrix} \cos\varphi_J + z'_n \sin\varphi_J & y'_n \cos\varphi_J & z'_n \cos\varphi_J - \sin\varphi_J \\ -y'_n & 1 & 0 \\ \sin\varphi_J - z'_n \cos\varphi_J & y'_n \sin\varphi_J & z'_n \sin\varphi_J + \cos\varphi_J \end{pmatrix}\Bigg|_{\xi_n = l_n} \cdot \begin{pmatrix} e_1 \\ e_2 \\ e_3 \end{pmatrix}.$$
$$(3.65)$$

Here we keep only the linear terms with respect to elastic deformations.

Let us obtain the kinetic and potential energy of Σ assuming that the center of mass of Σ is on the $O_2 \gamma_1$-axis and using formulas (3.57), (3.58), (3.61), and (3.65):

$$\begin{aligned} 2T_\Sigma = &\, J_1 \left\{ \dot{y}'_n \sin\varphi_J + y'_n \dot{\varphi}_R \cos\varphi_J + \dot{\varphi}_T (z'_n \cos(\varphi_R - \varphi_J) + \sin(\varphi_R - \varphi_J)) \right\}^2 \\ &+ J_2 \left\{ \dot{\varphi}_J - \dot{\varphi}_R + y'_n \dot{\varphi}_T \sin\varphi_R - \dot{z}'_n \right\}^2 \\ &+ J_3 \left\{ \dot{y}'_n \cos\varphi_J - y'_n \dot{\varphi}_R \sin\varphi_J + \dot{\varphi}_T (z'_n \sin(\varphi_R - \varphi_J) - \cos(\varphi_R - \varphi_J)) \right\}^2 \\ &+ m \left\{ c(\dot{z}'_n - \dot{\varphi}_J) \sin\varphi_J + \dot{l} + \dot{\varphi}_R (l z'_n - z_n + c \sin\varphi_J) \right. \\ &\left. + \dot{\varphi}_T (y_n \cos\varphi_R + y'_n (R - l \cos\varphi_R - c \sin\varphi_R \sin\varphi_J)) \right\}^2 \\ &+ m \left\{ \dot{y}_n + c\dot{y}'_n \cos\varphi_J - cy'_n \dot{\varphi}_R \sin\varphi_J \right. \\ &\left. + \dot{\varphi}_T (R - l \cos\varphi_R + z_n \sin\varphi_R - c \cos(\varphi_R - \varphi_J) + cz'_n \sin(\varphi_R - \varphi_J)) \right\}^2 \\ &+ m \left\{ \dot{z}_n + c(\dot{z}'_n - \dot{\varphi}_J) \cos\varphi_J + \dot{\varphi}_R (l + c \cos\varphi_J) - \dot{\varphi}_T \sin\varphi_R (y_n + cy'_n \cos\varphi_J) \right\}^2 \Bigg|_{\xi_n = l_n}, \end{aligned}$$
$$(3.66)$$

$$\Pi_\Sigma = mg \left\{ l \sin\varphi_R + z_n \cos\varphi_R + cz'_n \cos(\varphi_R - \varphi_J) + c \sin(\varphi_R - \varphi_J) \right\}\Big|_{\xi_n = l_n}.$$
$$(3.67)$$

Here J_1, J_2, J_3 are moments of inertia of the body Σ with respect to the axes passing through its center of mass and parallel to γ_1, γ_2, γ_3, respectively; m is the mass of Σ; c is the distance from O_2 to the center of mass of Σ. To obtain formula (3.66), we assume that the trihedron $O_2 \gamma_1 \gamma_2 \gamma_3$ corresponds to the principal axes of inertia of Σ.

3.5.2 Equations of Motion

Let us denote by $\tilde{C}^k[\alpha, \beta], k > 1$ the space of all continuously differentiable functions on $[\alpha, \beta]$ with piecewise-continuous derivatives up to the kth order. In order to derive the equations of motion for the mechanical system considered, we use Hamilton's principle with the Lagrange functional

$$L : \mathbb{R}^8 \times \tilde{C}^2[0, l_1] \times C^1[0, l_1] \times \cdots \times \tilde{C}^2[0, l_n] \times C^1[0, l_n] \to \mathbb{R}^1,$$

$$L(\varphi_R, \dot{\varphi}_R, \varphi_T, \dot{\varphi}_T, \varphi_J, \dot{\varphi}_J, l, \dot{l}, y_1, \dot{y}_1, \ldots, z_n, \dot{z}_n) = \sum_{j=0}^{n}(T_j - \Pi_j) + T_\Sigma - \Pi_\Sigma. \tag{3.68}$$

The terms appearing in formula (5.3) are given by expressions (3.62), (3.63), (3.64), (3.66), and (3.67). A justification of Hamilton's principle for problems of continuum mechanics is presented in the monograph by Berdichevsky [23].

If the the initial (at $t = t_1$) and the final (at $t = t_2$) states of the mechanical system are given, then the motion $\varphi_R(t), \varphi_T(t), \varphi_J(t), l(t), y_j(\cdot, t), z_j(\cdot, t), j = \overline{1, n}$ under the action of active forces on $t \in [t_1, t_2]$ satisfies Hamilton's principle as follows:

$$\delta \int_{t_1}^{t_2} L \, dt + \int_{t_1}^{t_2} (M_T(t)\delta\varphi_T(t) + M_R(t)\delta\varphi_R(t) + M_J(t)\delta\varphi_J(t) + F_l(t)\delta l(t)) \, dt = 0, \tag{3.69}$$

for all admissible variations

$$\delta\varphi_T, \delta\varphi_R, \delta\varphi_J, \delta l \in C^1[t_1, t_2], \quad \delta y_j, \delta z_j \in C^1\left([0, l_j] \times [t_1, t_2]\right), \quad j = \overline{1, n},$$

satisfying conditions

$$\delta\varphi_R(t_1) = \delta\varphi_R(t_2) = \delta\varphi_T(t_1) = \delta\varphi_T(t_2) = \delta\varphi_J(t_1) = \delta\varphi_J(t_2) = 0,$$
$$\delta l(t_1) = \delta l(t_2) = 0,$$
$$y_j(\xi_j, t_1) = y_j(\xi_j, t_2) = z_j(\xi_j, t_1) = z_j(\xi_j, t_2) = 0, \quad \forall \xi_j \in [0, l_j],$$
$$\delta y_1(0, t) = \delta y_1'(0, t) = \delta z_1(0, t) = \delta z_1'(0, t) = 0,$$
$$\delta y_j(\cdot, t), \delta z_j(\cdot, t) \in \tilde{C}^2[0, l_j], \forall t \in [t_1, t_2], \quad j = \overline{1, n}, \tag{3.70}$$

and the geometric contact condition of type (3.60). Besides, the functions $y_j(\xi_j, t)$, $z_j(\xi_j, t)$ of flexible displacements should satisfy the geometric boundary conditions

$$y_1(0, t) = y_1'(0, t) = z_1(0, t) = z_1'(0, t) = 0, \quad \forall t \in [t_1, t_2], \tag{3.71}$$

and contact condition (3.60).

In formula (3.69), the control torques M_R, M_T, M_J and the force l ensure rotations of system elements by angles $\varphi_R, \varphi_T, \varphi_J$, and the telescoping of links along the generalized coordinate L, respectively. In practice these control torques are implemented by hydraulic actuators, and the force F_L (rope tension) is regulated by the telescoping mechanism.

We denote

$$y(x, t) = y_j(x - a_j(l(t)), t), \ \ z(x, t) = z_j(x - a_j(l(t)), t), \ \ as \ x - a_j(l(t)) \in [0, l_j], \tag{3.72}$$

$$\rho(x; l) = \sum_{j=1}^{n} \chi_{[a_j(l), b_j(l)]} \rho_j(x - a_j(l)), \ \ \rho^*(x; l) = \sum_{j=2}^{n} \frac{j-1}{n-1} \chi_{[a_j(l), b_j(l)]} \rho_j(x - a_j(l)),$$

$$c_y(x; l) = \sum_{j=1}^{n} \chi_{[a_j(l), b_j(l)]} E_j(x - a_j(l)) \mathscr{I}_{jy}(x - a_j(l)),$$

$$c_z(x; l) = \sum_{j=1}^{n} \chi_{[a_j(l), b_j(l)]} E_j(x - a_j(l)) \mathscr{I}_{jz}(x - a_j(l)),$$

where $\chi_{[a_j, b_j]}$ is the indicator of the segment $[a_j, b_j]$, i.e.

$$\chi_{[a,b]}(x) = \begin{cases} 1, & x \in [a, b], \\ 0, & x \notin [a, b]. \end{cases}$$

Obviously, the functions $y(x, t)$ and $z(x, t)$ are well-defined for all $x \in [0, l]$ in (3.72) provided that all y_j, z_j satisfy condition (3.60).

We assume that $y_j(\cdot, t), z_j(\cdot, t) \in \tilde{C}^4[0, l_j], \forall t \in [t_1, t_2], j = \overline{1, n}$ in formula (3.69). Integrating by parts the expression in (3.69) and using (3.60), (3.70)–(3.72), we get the following system of equations with partial derivatives:

$$\ddot{y} + \frac{1}{\rho}(c_z y'')'' = \dot{\varphi}_T \left(y \dot{\varphi}_T - 2\dot{z} \sin \varphi_R - 2\dot{\varphi}_R (z \cos \varphi_R + x \sin \varphi_R) + \frac{2\rho^* \cos \varphi_R}{\rho} \right)$$
$$+ \ddot{\varphi}_T (x \cos \varphi_R - z \sin \varphi_R - R), \tag{3.73}$$

$$\ddot{z} + \frac{1}{\rho}(c_y z'')'' = y \ddot{\varphi}_T \sin \varphi_R - x \ddot{\varphi}_R - g \cos \varphi_R$$
$$+ \dot{\varphi}_T \sin \varphi_R (2\dot{y} + \dot{\varphi}_T (R + z \sin \varphi_R - x \cos \varphi_R)) \tag{3.74}$$
$$+ \dot{\varphi}_R \left(z \dot{\varphi}_R - \frac{2\rho^* i}{\rho} \right)$$

as $x \in [0, l]\backslash S$, $S = \{a_1(l), b_1(l), \ldots, a_n(l), b_n(l)\}$, and the boundary conditions

$$y|_{x=0} = z|_{x=0} = y'|_{x=0} = z'|_{x=0} = 0, \tag{3.75}$$

$$\lim_{x \to p-0} \frac{\partial^k (c_z y'')}{\partial x^k} = \lim_{x \to p+0} \frac{\partial^k (c_z y'')}{\partial x^k},$$

$$\lim_{x \to p-0} \frac{\partial^k (c_y z'')}{\partial x^k} = \lim_{x \to p+0} \frac{\partial^k (c_y z'')}{\partial x^k}, \quad \forall p \in (0, l) \cap S, \; k = 0, 1, \tag{3.76}$$

$$\frac{1}{m}(c_z y'')' - \ddot{y} - c\ddot{y}' \cos \varphi_J - \ddot{\varphi}_T (R - l \cos \varphi_R - c \cos(\varphi_R - \varphi_J))\bigg|_{x=l} + \cdots = 0, \tag{3.77}$$

$$\frac{1}{m}(c_y z'')' - g \cos \varphi_R - \ddot{z} + c(\ddot{\varphi}_J - \ddot{z}') \cos \varphi_J - \ddot{\varphi}_R (l + c \cos \varphi_J)\bigg|_{x=l} + \cdots = 0, \tag{3.78}$$

$$- c_z y'' + m(R - l \cos \varphi_R - c \sin \varphi_R \sin \varphi_J)\dot{\varphi}_T$$
$$\times \left\{ c(\dot{z}' - \dot{\varphi}_J) \sin \varphi_J + \dot{l} + \dot{\varphi}_R(lz' - z + c \sin \varphi_J) \right.$$
$$\left. + \dot{\varphi}_T (y \cos \varphi_R + y'(R - l \cos \varphi_R - c \sin \varphi_R \sin \varphi_J)) \right\}$$
$$- J_1 \sin \varphi_J \left\{ \ddot{y}' \sin \varphi_J + \ddot{\varphi}_T(z' \cos(\varphi_R - \varphi_J) + \sin(\varphi_R - \varphi_J)) \right\}$$
$$- J_3 \cos \varphi_J \left\{ \ddot{y}' \cos \varphi_J + \ddot{\varphi}_T(z' \sin(\varphi_R - \varphi_J) - \cos(\varphi_R - \varphi_J)) \right\}$$
$$+ mc \sin \varphi_J \left\{ \ddot{y} + c\ddot{y}' \cos \varphi_J + \ddot{\varphi}_T (R - l \cos \varphi_R - c \cos(\varphi_R - \varphi_J)) \right\}\bigg|_{x=l} + \cdots = 0, \tag{3.79}$$

$$- c_y z'' + (J_2 + mc^2)(\ddot{\varphi}_J - \ddot{\varphi}_R - \ddot{z}')$$
$$- mc \left\{ \ddot{l} \sin \varphi_J + (\ddot{z} + l\ddot{\varphi}_R) \cos \varphi_J + g \cos(\varphi_R - \varphi_J) \right\}\bigg|_{x=l} + \cdots = 0, \tag{3.80}$$

where the prime stands for the differentiation with respect to x. We do not take into account nonlinear terms with respect to the flexible deformations in equations (3.73)–(3.80) in order to simplify the model. Here, and in the sequel, we use dots to denote the terms of higher than the first order with respect to y, z and their derivatives.

We also derive the Lagrange equations from expression (3.69) with generalized coordinates φ_R, φ_T, φ_J, and l:

$$
\begin{aligned}
&\big\{ I_0 + I_1 \sin^2 \varphi_R + I_3 \cos^2 \varphi_R + J_1 \sin^2(\varphi_R - \varphi_J) + J_3 \cos^2(\varphi_R - \varphi_J) \\
&\quad + m_0(R - d \cos \varphi_R)^2 + m(R - l \cos \varphi_R - c \cos(\varphi_R - \varphi_J))^2 \big\} \\
&\quad \times \ddot{\varphi}_T + m(R - l \cos \varphi_R - c \cos(\varphi_R - \varphi_J)) \ddot{y} \\
&\quad + \big\{ J_1 \sin \varphi_J \sin(\varphi_R - \varphi_J) - J_3 \cos \varphi_J \cos(\varphi_R - \varphi_J) \\
&\qquad + mc(R - l \cos \varphi_R - c \cos(\varphi_R - \varphi_J)) \cos \varphi_J \big\} \ddot{y}'\big|_{x=l} \\
&\quad + \int_0^l (\ddot{y} + (R - x \cos \varphi_R) \ddot{\varphi}_T)(R - x \cos \varphi_R) \rho \, dx + \cdots = M_T, \qquad (3.81)
\end{aligned}
$$

$$
\begin{aligned}
&\left\{ I_2 + J_2 + m_0 d^2 + m(c^2 + l^2 + 2lc \cos \varphi_J) + \int_0^l x^2 \rho \, dx \right\} \ddot{\varphi}_R \\
&+ \int_0^l x \ddot{z} \rho \, dx + mc \ddot{l} \sin \varphi_J + (J_2 + mc(c + l \cos \varphi_J))(\ddot{z}' - \ddot{\varphi}_J) + m(l + c \cos \varphi_J) \ddot{z} \\
&+ g \bigg\{ \int_0^l (x \cos \varphi_R - z \sin \varphi_R) \rho \, dx + (m_0 d + ml) \cos \varphi_R - mz \sin \varphi_R \\
&\qquad + mc(\cos(\varphi_R - \varphi_J) - z' \sin(\varphi_R - \varphi_J)) \bigg\}\bigg|_{x=l} + \cdots = M_R, \qquad (3.82)
\end{aligned}
$$

$$
\begin{aligned}
&(J_2 + mc^2)(\ddot{\varphi}_J - \ddot{z}') - mc(\ddot{z} \cos \varphi_J + \ddot{l} \sin \varphi_J) - \{ J_2 + mc(c + l \cos \varphi_J) \} \ddot{\varphi}_R \\
&+ mgc(z' \sin(\varphi_R - \varphi_J) - \cos(\varphi_R - \varphi_J))\big|_{x=l} + \cdots = M_J, \qquad (3.83)
\end{aligned}
$$

$$
\begin{aligned}
&\left(m + \sum_{j=2}^{n} \frac{(j-1)^2}{(n-1)^2} m_j \right) \ddot{l} + mc(\ddot{z}'\big|_{x=l} + \ddot{\varphi}_R - \ddot{\varphi}_J) \sin \varphi_J \\
&- g \left(m + \sum_{j=2}^{n} \frac{j-1}{n-1} m_j \right) \sin \varphi_R + \cdots = F_l, \qquad (3.84)
\end{aligned}
$$

where m_j is the mass of the jth link.

Thus, we have obtained the nonlinear system of integro-differential equations (3.73)–(3.84) that describes the controlled motion of a flexible multilink manipulator with taking into account the telescoping of links.

Let us emphasize that networks of the Euler–Bernoulli and Timoshenko beams have been considered in several publications, for instance, in the monograph [1], where different junction conditions have been studied. A mathematical model for

two serially connected Euler–Bernoulli beams with distributed masses at the ends
and a rigid rotating hub has been considered in the paper [32].

3.5.3 Equilibrium Conditions

An equilibrium position with constant values of the angles and the l coordinate is
one of basic working regimes of the manipulator. To ensure such an equilibrium,
one should compensate the gravity and elastic forces by control torques M_T, M_R,
M_J and the force F_l. For this purpose, let us find conditions for the existence of a
solution of the equations of motion in the following form:

$$\varphi_T = \text{const}, \ \ \varphi_R = \text{const}, \ \ \varphi_J = \text{const}, \ \ l = \text{const}, \ \ y(x,t) = 0, \ \ z(x,t) = z_0(x).$$
$$(3.85)$$

By substituting (3.85) into (3.73)–(3.84), we get

$$(c_y(x;l)z_0''(x))'' = -\rho g \cos \varphi_R, \ \ x \in [0,l] \backslash S, \ \ S = \{a_1(l), b_1(l), \ldots, a_n(l), b_n(l)\},$$
$$(3.86)$$

$$z_0(0) = z_0'(0) = 0, \ \ \lim_{x \to p-0} \frac{d^k(c_y z_0'')}{dx^k} = \lim_{x \to p+0} \frac{d^k(c_y z_0'')}{dx^k}, \ \ \forall p \in (0,l) \cap S, \ k = 0, 1,$$
$$(3.87)$$

$$(c_y z_0'')'|_{x=l} = mg \cos \varphi_R, \ \ c_y z_0''|_{x=l} = -mgc \cos(\varphi_R - \varphi_J), \qquad (3.88)$$

$$\frac{M_R}{g} = (m_0 d + ml) \cos \varphi_R - m z_0(l) \sin \varphi_R$$

$$+ \int_0^l (x \cos \varphi_R - z_0(x) \sin \varphi_R) \rho \, dx + mc(\cos(\varphi_R - \varphi_J) - z_0'(l) \sin(\varphi_R - \varphi_J)),$$
$$(3.89)$$

$$M_T = 0, \ \ M_J = mgc(z_0'(l) \sin(\varphi_R - \varphi_J) - \cos(\varphi_R - \varphi_J)),$$

$$F_l = -g \left(m + \sum_{j=2}^n \frac{j-1}{n-1} m_j \right) \sin \varphi_R. \qquad (3.90)$$

For a particular case $n = 1$, $\rho = const$, $c_y = const$, the solution of the boundary
problem (3.86)–(3.88) takes the form

$$z_0(x) = -\frac{gx^2 \cos \varphi_R}{24c_y} (\rho x^2 - 6mx + 12ml) - \frac{mgcx^2 \cos(\varphi_R - \varphi_J)}{2c_y}.$$

Then the reference torques and force may be defined by substituting $z_0(x)$ into (3.89)–(3.90).

3.5.4 Motion of a Single-Link Manipulator

Let us consider the equations of motion (3.73)–(3.84) for a particular case of a single-link manipulator ($n = 1$) [33]. Suppose also that the rigid body Σ is attached at its center of mass ($c = 0$) and does not perform relative rotations at the end of the manipulator ($\varphi_J = 0$).

Thus, the transverse displacement of the beam at a point $x \in [0, l]$ and time t is defined by functions $y(x, t)$ and $z(x, t)$, where l is the beam length. Assume that the equilibrium of type (3.85) is achieved for given values of the angle φ_R^0 and torque $M_R = M_R^0$, i.e.

$$\varphi_T(t) = 0, \quad \varphi_R(t) = \varphi_R^0, \quad y(x, t) = 0, \quad z(x, t) = z_0(x).$$

Then we write the system of linear approximation for equations (3.73)–(3.84) in a neighborhood of that equilibrium:

$$\ddot{y}(x, t) + \frac{1}{\rho}\left(c_z y''(x, t)\right)'' = \psi_T(x)\ddot{\varphi}_T, \quad x \in (0, l), \tag{3.91}$$

$$\ddot{z}(x, t) + \frac{1}{\rho}\left(c_y \tilde{z}''(x, t)\right)'' = g\tilde{\varphi}_R \sin \varphi_R^0 - x\ddot{\varphi}_R, \tag{3.92}$$

$$y|_{x=0} = \tilde{z}|_{x=0} = 0, \quad y'|_{x=0} = \tilde{z}'|_{x=0} = 0, \tag{3.93}$$

$$\frac{1}{m}(c_z y'')' - \ddot{y} + \psi_T(x)\ddot{\varphi}_T \bigg|_{x=l} = 0, \tag{3.94}$$

$$-c_z y'' - J_3 \ddot{y}' + J_3 \psi_T'(x)\ddot{\varphi}_T \big|_{x=l} = 0, \tag{3.95}$$

$$\frac{1}{m}(c_y \tilde{z}'')' + g\tilde{\varphi}_R \sin \varphi_R^0 - \ddot{z} - l\ddot{\varphi}_R \bigg|_{x=l} = 0, \tag{3.96}$$

$$c_y \tilde{z}'' + J_2(\ddot{\varphi}_R + \ddot{z}') \big|_{x=l} = 0, \tag{3.97}$$

$$\left\{ I_0 + (I_1 + J_1)\sin^2\varphi_R^0 + (I_3 + J_3)\cos^2\varphi_R^0 + m_0(R - d\cos\varphi_R^0)^2 \right.$$

$$\left. + m(R - l\cos\varphi_R^0)^2 + \int_0^l (R - x\cos\varphi_R^0)^2\rho\, dx \right\} \ddot{\varphi}_T$$

$$+ \int_0^l (R - x\cos\varphi_R^0)\ddot{y}\rho\, dx + \left\{ mR\ddot{y} - (ml\ddot{y} + J_3\ddot{y}')\cos\varphi_R^0 \right\}\Big|_{x=l} = M_T,$$

$$(3.98)$$

$$\left\{ I_2 + J_2 + m_0 d^2 + ml^2 + \int_0^l x^2\rho\, dx \right\} \ddot{\tilde{\varphi}}_R + \int_0^l \ddot{\tilde{z}}x\rho\, dx + \left\{ ml\ddot{\tilde{z}} + J_2\ddot{\tilde{z}}' \right\}\Big|_{x=l}$$

$$- g\left\{ \left[\int_0^l \ddot{\tilde{z}}\rho\, dx + m\tilde{z}|_{x=l} + \left(m_0 d + ml + \int_0^l x\rho\, dx \right)\tilde{\varphi}_R \right] \sin\varphi_R^0 \right.$$

$$- g\left\{ \left[\int_0^l z_0\rho\, dx + mz_0(l) \right\} \tilde{\varphi}_R\cos\varphi_R^0 = \tilde{M}_R, \right. \qquad (3.99)$$

where

$$\tilde{z}(x,t) = z(x,t) - z_0(x), \quad \tilde{\varphi}_R(t) = \varphi_R(t) - \varphi_R^0, \quad \tilde{M}_R = M_R - M_R^0,$$
$$\psi_T(x) = x\cos\varphi_R^0 - z_0(x)\sin\varphi_R^0 - R.$$

To simplify these equations, we substitute expressions (3.91), (3.92), (3.94)–(3.97) for $\ddot{y}(x,t), \ddot{\tilde{z}}(x,t), \ddot{y}, \ddot{y}', \ddot{\tilde{z}}, \ddot{\tilde{z}}'\big|_{x=l}$ into Eqs (3.98) and (3.99) and perform the integration by parts using boundary conditions (3.93). Then Eqs (3.98) and (3.99) take the form

$$\ddot{\varphi}_T = u_T, \quad \ddot{\tilde{\varphi}}_R = u_R, \qquad (3.100)$$

where

$$u_T = \{ I_0 + (I_1 + J_1)\sin^2\varphi_R^0 + m_0(R - d\cos\varphi_R^0)^2$$
$$+ (I_3\cos\varphi_R^0 + J_3 z_0'(l)\sin\varphi_R^0)\cos\varphi_R^0$$

$$+ \left(m(l\cos\varphi_R^0 - R)z_0(l) + \int_0^l (x\cos\varphi_R^0 - R)z_0\rho\, dx \right)\sin\varphi_R^0 \}^{-1}$$

$$\times \left\{ M_T - \left(R(c_z y'')' + c_z y''\cos\varphi_R^0 \right)\Big|_{x=0} \right\}, \qquad (3.101)$$

$$u_R = \{I_2 + m_0 d^2\}^{-1} \times \left\{ \tilde{M}_R + c_y \, \tilde{z}''|_{x=0} + g \left(\int_0^l \tilde{z} \rho \, dx + m \, \tilde{z}|_{x=l} + m_0 d \right) \sin \varphi_R^0 \right.$$

$$\left. + g \left(\int_0^l z_0 \rho \, dx + m z_0(l) \right) \tilde{\varphi}_R \cos \varphi_R^0 \right\}. \qquad (3.102)$$

For each $\tilde{\varphi}_R(t)$, $y(\cdot, t)$, and $\tilde{z}(\cdot, t)$, formulas (3.101) and (3.102) define a one-to-one correspondence between the torques $(M_T, \tilde{M}_R)$ and angular accelerations (u_T, u_R). Thus, we will refer to $(u_T, u_R) \in \mathbb{R}^2$ as to the control in linear system (3.91)–(3.97), (3.100).

3.5.5 Stability of the Linearized System

In order to represent system (3.91)–(3.97) in the operator form, we consider the linear space

$$X = \left\{ \begin{pmatrix} \eta \\ \zeta \\ \phi \\ \omega \\ p \\ q \end{pmatrix} : \eta \in H^2(0, l), \, \zeta \in L^2(0, l), \, \eta(0) = \eta'(0) = 0, \, \phi, \omega, p, q \in \mathbb{R} \right\},$$

where $H^k(0, l)$ is the Sobolev space of functions from $L^2(0, l)$ with generalized derivatives up to the k-th order from $L^2(0, l)$ (see, e.g., [34]). The inner product for elements

$$\xi_1 = (\eta_1, \zeta_1, \phi_1, \omega_1, p_1, q_1)^T \in X, \quad \xi_2 = (\eta_2, \zeta_2, \phi_2, \omega_2, p_2, q_2)^T \in X,$$

is defined by the formula

$$\langle \xi_1, \xi_2 \rangle_X = \int_0^l (\eta_1''(x) \eta_2''(x) + \zeta_1(x) \zeta_2(x)) \, dx + \phi_1 \phi_2 + \omega_1 \omega_2 + p_1 p_2 + q_1 q_2.$$

It is easy to show by using the Cauchy–Schwartz inequality that each function $\eta \in H^1(0, l)$ such that $\eta(0) = 0$ satisfies the following Friedrichs–Wirtinger type inequality (cf. [35]):

$$\|\eta\|_{L^2(0,l)}^2 = \int_0^l \eta^2(x)\,dx = \int_0^l \left(\int_0^x \eta'(s)\,dx\right)^2 dx$$

$$\leq \int_0^l \left(\int_0^x ds \cdot \int_0^x \eta'^2(s)\,ds\right) dx$$

$$= \int_0^l \left(\eta'^2(s)\int_s^l x\,dx\right) ds \leq \frac{l^2}{2}\int_0^l \eta'^2(s)\,ds = \frac{l^2}{2}\|\eta'\|_{L^2(0,l)}^2. \quad (3.103)$$

Similarly, for $\eta \in H^2(0,l)$, $\eta'(0) = 0$ we get

$$\|\eta'\|_{L^2(0,l)}^2 \leq \frac{l^2}{2}\|\eta''\|_{L^2(0,l)}^2. \quad (3.104)$$

Inequalities (3.103)–(3.104) imply the following estimate for each function $\eta \in H^2(0,l)$, $\eta(0) = \eta'(0) = 0$:

$$\|\eta\|_{H^2(0,l)} = \left(\|\eta\|_{L^2}^2 + \|\eta'\|_{L^2}^2 + \|\eta''\|_{L^2}^2\right)^{1/2} \leq \left(\frac{l^4}{4} + \frac{l^2}{2} + 1\right)^{1/2}\|\eta''\|_{L^2(0,l)}.$$

Hence, it is easy to see that the norm $\|\xi\|_X = \sqrt{\langle\xi,\xi\rangle_X}$ is equivalent to the standard norm in a linear subspace X of the Hilbert space $H^2(0,l) \times L^2(0,l) \times \mathbb{R}^4$, therefore $(X, \|\cdot\|_X)$ is the Hilbert space.

Let us define a linear unbounded operator $A : D(A) \to X$ and an element $B \in X$ as follows:

$$A : \xi = \begin{pmatrix} \eta \\ \zeta \\ \phi \\ \omega \\ p \\ q \end{pmatrix} \mapsto A\xi = \begin{pmatrix} \zeta \\ -\frac{1}{\rho}(c\eta'')'' + \gamma\phi \\ \omega \\ 0 \\ \gamma\phi + \frac{1}{m}(c\eta'')'|_{x=l} \\ -\frac{c}{J}\eta''|_{x=l} \end{pmatrix}, \quad B = \begin{pmatrix} 0 \\ \psi \\ 0 \\ 1 \\ \psi(l) \\ \psi'(l) \end{pmatrix}, \quad (3.105)$$

where the domain of definition of A is

$$D(A) = \left\{\xi \in X : \begin{array}{l} \eta \in H^4(0,l), \zeta \in H^2(0,l), \\ \zeta(0) = \zeta'(0) = 0, \\ p = \zeta(l), q = \zeta'(l) \end{array}\right\}. \quad (3.106)$$

We assume that $c(x) > 0$ and $\psi(x)$ are $C^2[0,l]$-functions, and that $J > 0$, γ are constants.

Let $\left(y(x,t), \tilde{z}(x,t), \varphi_T(t), \tilde{\varphi}_R(t)\right)$ be a classical solution of the boundary value problem (3.91)–(3.97), (3.100) with the control $(u_T(t), u_R(t))$ for $0 \leq t < \tau$, $\tau \leq +\infty$. We define

$$\xi_T(t) = \begin{pmatrix} y(\cdot, t) \\ \dot{y}(\cdot, t) \\ \varphi_T(t) \\ \dot{\varphi}_T(t) \\ \dot{y}(l, t) \\ \dot{y}'(l, t) \end{pmatrix}, \quad \xi_R(t) = \begin{pmatrix} \tilde{z}(\cdot, t) \\ \dot{\tilde{z}}(\cdot, t) \\ \tilde{\varphi}_R(t) \\ \dot{\tilde{\varphi}}_R(t) \\ \dot{\tilde{z}}(l, t) \\ \dot{\tilde{z}}'(l, t) \end{pmatrix}, \tag{3.107}$$

so that $\xi_T(t) \in D(A)$ and $\xi_R(t) \in D(A)$ for all $t \in [0, \tau)$. Consider a pair (A_T, B_T) obtained from (A, B) by the substitution of

$$. \quad \psi(x) = \psi_T(x), \quad c(x) = c_z(x), \quad J = J_3, \quad \gamma = 0$$

into formula (3.105). Similarly, let us define a pair (A_R, B_R) by formulas for (A, B) with

$$\psi(x) = -x, \quad c(x) = c_y(x), \quad J = J_2, \quad \gamma = g \sin \varphi_R^0.$$

Then the problem (3.91)–(3.97), (3.100) is reduced to the following system:

$$\dot{\xi}_T = A_T \xi_T + B_T u_T, \tag{3.108}$$

$$\dot{\xi}_R = A_R \xi_R + B_R u_R, \tag{3.109}$$

where (ξ_T, ξ_R) is the state vector, and (u_T, u_R) is the control.

Thus, we have constructed an operator form of the problem (3.91)–(3.97) and (3.100) with $\xi_T, \xi_R \in X$ and $u_T, u_R \in \mathbb{R}$. As we see, system (3.108) and (3.109) is divided into two subsystems, therefore, the stabilization problem can be considered separately for components ξ_T and ξ_R.

Theorem 3.3 [33] *Let us consider the following Cauchy problem on $t \geq 0$:*

$$\dot{\xi}(t) = A\xi(t) + Bu, \tag{3.110}$$

$$\xi(0) = \xi_0 \in X, \tag{3.111}$$

where A and B are defined by formulas (3.105), $u = h(\xi)$,

$$h(\xi) = -\frac{1}{\beta} \left\{ k\omega + \left(\alpha - \gamma \left(\int_0^l \rho\psi \, dx + m\psi(l) \right) \right) \phi \right.$$

$$+ \int_0^l c\eta''\psi'' dx + \left(c\eta''\psi' - (c\eta'')'\psi \right)\Big|_{x=0} - \gamma \left(\int_0^l \rho\eta \, dx + m\eta(l) \right) \right\},$$

$$\tag{3.112}$$

*where $k > 0$ is an arbitrary constant, and constants $\alpha > 0$, $\beta > 0$ are large
enough.*

*Then the Cauchy problem (3.110), (3.111) with control (3.112) is well-posed on
$t \geq 0$, and its solution $\xi = 0$ is strongly stable in the sense of Lyapunov, that is, for
any $\varepsilon > 0$ there is a $\delta = \delta(\varepsilon) > 0$ such that each solution of problem (3.110)–(3.112)
satisfies the condition*

$$\|\xi_0\| < \delta \Rightarrow \|\xi(t)\| < \varepsilon, \quad \forall t \geq 0.$$

Proof Let us consider the following quadratic functional on X:

$$2V(\xi) = \alpha\phi^2 + \beta\omega^2 + \int_0^l \left\{ (\zeta - \psi\omega)^2\rho + \eta''^2 c \right\} dx$$

$$+ m\left\{ p - \psi(l)\omega \right\}^2 + J\left\{ q - \psi'(l)\omega \right\}^2$$

$$- 2\gamma\phi \left\{ \int_0^l \eta\rho \, dx + m\eta(l) \right\}. \tag{3.113}$$

The time derivative of functional V along the trajectories of system (3.110) for
$\xi \in D(A)$ is

$$\dot{V}(\xi) = \langle \nabla V(\xi), A\xi + Bu \rangle$$

$$= \int_0^l \left(c\zeta''\eta'' - \zeta \cdot (c\eta'')'' \right) dx + \left(p\,(c\eta'')' - qc\eta'' \right)\Big|_l$$

$$+ \left\{ \alpha\phi + \beta u + \int_0^l \left(\psi \cdot (c\eta'')'' - \rho\gamma(\phi\psi + \eta) \right) dx \right\}$$

$$+ \left(c\psi'\eta'' - \psi \cdot (c\eta'')' - m\gamma(\phi\psi + \eta) \right)\Big|_{x=l} \omega. \tag{3.114}$$

Integrating by parts and applying boundary conditions (3.106) we have

$$\int_0^l \zeta \cdot (c\eta'')'' dx = \zeta \cdot (c\eta'')'\Big|_{x=0}^l - \int_0^l \zeta' \cdot (c\eta'')' dx$$

$$= \left(\zeta \cdot (c\eta'')' - \zeta'c\eta'' \right)\Big|_{x=l} + \int_0^l \zeta''c\eta'' dx,$$

$$\int_0^l \psi \cdot (c\eta'')'' dx = \psi \cdot (c\eta'')' \Big|_{x=0}^l - \int_0^l \psi'(c\eta'')' dx$$

$$= \left(\psi \cdot (c\eta'')' - \psi' c\eta'' \right) \Big|_{x=0}^l + \int_0^l \psi'' c\eta_T'' dx.$$

By substituting the above formulas into (3.114) and using boundary conditions $p = \zeta(l), q = \zeta'(l)$ from (3.106), we obtain the following expression for $\dot{V}$:

$$\dot{V}(\xi) = \left\{ \left(\alpha - \gamma \left(\int_0^l \rho\psi \, dx + m\psi(l) \right) \right) \phi + \beta u \right.$$

$$+ \int_0^l c\psi'' \eta'' dx + \left(c\psi' \eta'' - \psi \cdot (c\eta'')' \right) \Big|_{x=0}^l - \gamma \left(\int_0^l \rho\eta \, dx + m\eta(l) \right) \right\} \omega.$$

$$(3.115)$$

For u defined by expression (3.112), formula (3.115) implies

$$\dot{V}(\xi) = -k\omega^2 \le 0, \quad (k = const > 0). \tag{3.116}$$

Let us show that the functional $V(\xi)$ satisfy estimates

$$M_1 \|\xi\|_X^2 \le 2V(\xi) \le M_2 \|\xi\|_X^2 \tag{3.117}$$

with some constants $0 < M_1 \le M_2 < +\infty$. On the one hand, by applying inequalities $(a+b)^2 \le 2a^2 + 2b^2$ and $2ab \le a^2 + b^2$ to (3.113), we get

$$2V(\xi) \le \alpha\phi^2 + \beta\omega^2 + \int_0^l \left(c\eta''^2 + 2(\zeta^2 + \psi^2\omega^2)\rho \right) dx$$

$$+ 2m \left(p^2 + \psi^2(l)\omega^2 \right) + 2J \left(q^2 + \psi'^2(l)\omega^2 \right)$$

$$+ \gamma^2\phi^2 + 2 \left(\int_0^l \eta\rho \, dx \right)^2 + 2 (m\eta(l))^2. \tag{3.118}$$

Then the Cauchy–Schwartz inequality implies

$$\left(\int_0^l \eta\rho \, dx \right)^2 \le \int_0^l \eta^2 \, dx \int_0^l \rho^2 \, dx, \tag{3.119}$$

$$\eta^2(l) = \left(\int_0^l \eta' \, dx \right)^2 \leq \int_0^l dx \int_0^l \eta'^2 dx. \tag{3.120}$$

Functions $\eta(x)$ and $\eta'(x)$ with boundary conditions $\eta(0) = \eta'(0) = 0$ satisfy the Friedrichs–Wirtinger inequalities of the following form (see [35, 36]):

$$\int_0^l \eta^2 dx \leq \frac{l^2}{2} \int_0^l \eta'^2 dx \leq \frac{l^4}{4} \int_0^l \eta''^2 dx. \tag{3.121}$$

Inequalities (3.119)–(3.121) lead to

$$\left(\int_0^l \eta\rho \, dx \right)^2 + m^2\eta^2(l) \leq \int_0^l \eta^2 \, dx \int_0^l \rho^2 \, dx$$

$$+ lm^2 \int_0^l \eta'^2 dx \leq \frac{l^3}{2} \left(m^2 + \frac{l}{2} \int_0^l \rho^2 dx \right) \int_0^l \eta''^2 dx. \tag{3.122}$$

Applying this inequality to (3.118), we get an estimate $2V(\xi) \leq M_2 \|\xi_T\|_X^2$ with

$$M_2 = \max \left\{ \alpha + \gamma^2, 2m, 2J, 2 \max_{x\in[0,l]} \rho(x), \right.$$

$$\beta + 2 \int_0^l \psi^2 \rho \, dx + 2J\psi'^2(l) + 2m\psi^2(l),$$

$$\left. l^3 \left(m^2 + \frac{l}{2} \int_0^l \rho^2 dx \right) + \max_{x\in[0,l]} c(x) \right\}.$$

On the other hand, inequality $a^2 = (a-b+b)^2 \leq 2(a-b)^2 + 2b^2$ implies $(a-b)^2 \geq a^2/2 - b^2$. Using the latter inequality together with $-2ab \geq -\varkappa^2 a^2 - b^2/\varkappa^2$ ($\varkappa \neq 0$) in (3.113), we conclude that

$$2V(\xi) \geq \alpha\phi^2 + \beta\omega^2 + \int_0^l \left(c\eta''^2 + \frac{\rho}{2}\zeta^2 - \rho\psi^2\omega^2 \right) dx$$

$$+ m\left(\frac{p^2}{2} - \psi^2(l)\omega^2 \right) + J\left(\frac{q^2}{2} - \psi'^2(l)\omega^2 \right)$$

$$- \varkappa^2 \gamma^2 \phi^2 - \frac{1}{\varkappa^2} \left(\int_0^l \eta \rho \, dx + m \eta(l) \right)^2$$

$$\geq \left(\alpha - \varkappa^2 \gamma^2 \right) \phi^2 + \frac{m}{2} p^2 + \frac{J}{2} q^2 + \frac{1}{2} \int_0^l \zeta^2 \rho \, dx$$

$$+ \left(\beta - \int_0^l \rho \psi^2 \, dx - m \psi^2(l) - J \psi'^2(l) \right) \omega^2$$

$$+ \left\{ \min_{[0,l]} c - \frac{l^3}{\varkappa^2} \left(m^2 + \frac{l}{2} \int_0^l \rho^2 dx \right) \right\} \int_0^l \eta''^2 dx. \qquad (3.123)$$

Here we have also used inequality (3.122). Estimation (3.123) implies that $2V(\xi) \geq M_1 \|\xi\|_X^2$, where

$$M_1 = \min \left\{ \alpha - \varkappa^2 \gamma^2, \frac{m}{2}, \frac{J}{2}, \frac{1}{2} \min_{x \in [0,l]} \rho(x), \right.$$

$$\beta - \int_0^l \rho \psi^2 \, dx - m \psi^2(l) - J \psi'^2(l),$$

$$\left. \min_{x \in [0,l]} c(x) - \frac{l^3}{\varkappa^2} \left(m^2 + \frac{l}{2} \int_0^l \rho^2 dx \right) \right\} > 0$$

provided that

$$\varkappa^2 > \frac{l^3}{\min_{x \in [0,l]} c(x)} \left(m^2 + \frac{l}{2} \int_0^l \rho^2 dx \right),$$

$$\alpha > \varkappa^2 \gamma^2, \ \beta > \int_0^l \rho \psi^2 \, dx + m \psi^2(l) + J \psi'^2(l).$$

In the rest of the section, we assume that constants α, β, and $\varkappa$ satisfy the above inequalities.

Is follows from the estimate (3.117) that the norms $\|\xi\|_X$ and $\|\xi\|_V = \sqrt{V(\xi)}$ are equivalent in X. Let us rewrite system (3.110) with the feedback control u (3.112) as $\dot{\xi} = \tilde{A}\xi$, where the domain of definition $D(\tilde{A}) = D(A)$ is dense in X. Inequality (3.116) implies that the operator $\tilde{A}$ is dissipative in X with the norm $\|\cdot\|_V$. Then $\tilde{A}$ is the infinitesimal generator of a C_0-semigroup of contractions $\{e^{t\tilde{A}}\}_{t \geq 0}$ on X with respect to the norm $\| \cdot \|_V$ by the Lumer–Phillips theorem (Theorem 1.3). Hence, the Cauchy problem (3.110)–(3.112) has a unique mild solution $\xi(t) = e^{t\tilde{A}}\xi_0$ on $t \geq 0$ for any $\xi_0 \in X$, and such solution is classical if $\xi_0 \in D(A)$. Since $\{e^{t\tilde{A}}\}_{t \geq 0}$ is a semigroup of contractions (under equivalent renormalization in X), we conclude that

$$\| \xi(t) \|_V \leq \| \xi_0 \|_V, \quad \forall t \geq 0.$$

This fact and the estimate (3.117) imply that

$$\| \xi(t) \|_X^2 \leq \frac{2V(\xi(t))}{M_1} \leq \frac{2V(\xi_0)}{M_1} \leq \frac{M_2}{M_1} \| \xi_0 \|_X^2.$$

The inequality obtained proves the strong stability of the solution $\xi = 0$ in the sense of Lyapunov, and we may use $\delta(\varepsilon) = \varepsilon \sqrt{M_1/M_2}$ in the stability definition.

Remark 3.1 Let us write the controls obtained by applying formula (3.112) to subsystems (3.108) and (3.109) and using representation (3.107) separately:

$$u_T = -\frac{1}{\beta}\left\{ \alpha\varphi_T + k\dot{\varphi}_T + \int_0^l c_z y'' \psi'' dx + \left(c_z y'' \psi' - (c_z y'')' \psi \right)\Big|_{x=0} \right\},$$

$$u_R = -\frac{1}{\beta}\left\{ \alpha\tilde{\varphi}_R + k\dot{\tilde{\varphi}}_R - c_y \tilde{z}''\big|_{x=0} \right.$$

$$\left. + g\left(\int_0^l (x\tilde{\varphi}_R - \tilde{z})\rho \, dx + m(l\tilde{\varphi}_R - \tilde{z}|_{x=l}) \right) \sin\varphi_R^0 \right\}.$$

To implement these controls, one has to compute u_T and u_R depending on the measurements of $\varphi_T, \tilde{\varphi}_R, \dot{\varphi}_T, \dot{\tilde{\varphi}}_R, y, \tilde{z}$, and then apply formulas (3.101) and (3.102) to generate the control torques M_T and M_R. An advantage of this approach is that the information about time derivatives of $y(x, t)$ and $\tilde{z}(x, t)$ is not needed for the control design.

3.5.6 Localization of the Limit Set

Theorem 3.3 states the well-posedness of the abstract Cauchy problem and non-asymptotic stability of the trivial solution of the closed-loop system. However, as the time derivative of the Lyapunov functional $V(\xi)$ is not negative definite, the limit behavior of trajectories as $t \to +\infty$ requires further study.

Consider the linear operator

$$\tilde{A} : \xi \mapsto A\xi + Bh(\xi),$$

where A, B, and $u = h(\xi)$ are defined by formulas (3.105) and (3.112). Then differential equation (3.110) with the control $u = h(\xi)$ reads as follows:

$$\frac{d}{dt}\xi(t) = \tilde{A}\xi(t), \quad t \geq 0, \tag{3.124}$$

Since $h(\xi)$ is defined for all $\xi \in D(A)$ in (3.112), it is easy to see that $D(\tilde{A}) = D(A)$. In the sequel, we assume that γ is a real constant, α, β, c, ρ, m, J are positive constants, and $\psi(x)$ is a function of class $C^2[0, l]$.

Theorem 3.3 implies that the unbounded operator $\tilde{A} : D(\tilde{A}) \to X$ is the infinitesimal generator of a strongly continuous semigroup of linear operators $\{e^{t\tilde{A}}\}_{t \geq 0}$ on X, provided that α and β are large enough. In addition, the function $V(\xi(t))$ is non-increasing along the solutions of Eq. (3.110) with the feedback law $u = h(\xi)$ due to inequality $\dot{V} \leq 0$. Thus, there is a natural question about the possibility of studying the limit behavior of trajectories by using a modification of the invariance principle. According to conditions of Theorem 2.4, the positive answer to this question requires the compactness analysis of trajectories in X. Let us formulate the basic result about the precompactness of trajectories for differential equation (3.124).

Theorem 3.4 *Suppose that constants α and β are large enough, and that $k > 0$. Then each trajectory $\{\xi(t)\}_{t \geq 0}$ of differential equation (3.110) with the feedback control $u = h(\xi)$ defined by (3.112) is contained in a compact subset of X.*

Proof Consider the following equation for $\xi \in D(A)$ with parameters $\lambda > 0$, $u \in \mathbb{R}$, $\tilde{\xi} \in X$:

$$A\xi + Bu - \lambda\xi = \tilde{\xi}. \tag{3.125}$$

We rewrite this equations for the components of vectors $\xi = (\eta, \zeta, \phi, \omega, p, q)^T$ and $\tilde{\xi} = (\tilde{\eta}, \tilde{\zeta}, \tilde{\phi}, \tilde{\omega}, \tilde{p}, \tilde{q})^T$:

$$\zeta(x) - \lambda\eta(x) = \tilde{\eta}(x), \quad -\frac{c}{\rho}\frac{d^4\eta(x)}{dx^4} + \gamma\phi - \lambda\zeta(x) + \psi(x)u = \tilde{\zeta}(x), \quad x \in (0, l),$$
$$\tag{3.126}$$

$$\omega - \lambda\phi = \tilde{\phi}, \quad -\lambda\omega + u = \tilde{\omega}, \tag{3.127}$$

$$\gamma\phi + \frac{c}{m}\eta'''(l) - \lambda p + \psi(l)u = \tilde{p}, \quad -\frac{c}{J}\eta''(l) - \lambda q + \psi'(l)u = \tilde{q}. \tag{3.128}$$

By solving Eq. (3.127) with respect to ϕ and ω, we get

$$\phi = \frac{u - \tilde{\omega}}{\lambda^2} - \frac{\tilde{\phi}}{\lambda}, \quad \omega = \frac{u - \tilde{\omega}}{\lambda}. \tag{3.129}$$

Let us express $\zeta(x)$ from the first equation of (3.126):

$$\zeta(x) = \tilde{\eta}(x) + \lambda\eta(x), \tag{3.130}$$

and substitute representations (3.129) and (3.130) into the second equation of (3.126). Then we get the differential equation with respect to $\eta(x)$:

$$\frac{d^4\eta(x)}{dx^4} + \frac{\lambda^2\rho}{c}\eta(x) = f_0(x)u + f_1(x,\tilde{\xi}), \quad x \in (0,l), \tag{3.131}$$

where

$$f_0(x) = \frac{\rho}{c}\left(\frac{\gamma}{\lambda^2} + \psi(x)\right), \quad f_1(x,\tilde{\xi}) = -\frac{\rho}{c}\left(\frac{\gamma}{\lambda}\tilde{\phi} + \frac{\gamma}{\lambda^2}\tilde{\omega} + \lambda\tilde{\eta}(x) + \tilde{\zeta}(x)\right).$$

The property $\xi \in D(A)$ implies boundary conditions

$$\eta(0) = \eta'(0) = 0, \tag{3.132}$$

and $p = \zeta(l), q = \zeta'(l)$. By using (3.130), the last two conditions can be represented as

$$p = \tilde{\eta}(l) + \lambda\eta(l), \quad q = \tilde{\eta}'(l) + \lambda\eta'(l). \tag{3.133}$$

Substituting equalities (3.129) and (3.133) into (3.128), we get the following boundary conditions for $\eta(x)$ at $x = l$:

$$\eta'''(l) - \frac{\lambda^2 m}{c}\eta(l) = g_{10}u + g_{11}(\tilde{\xi}), \tag{3.134}$$

$$\eta''(l) + \frac{\lambda^2 J}{c}\eta'(l) = g_{20}u + g_{21}(\tilde{\xi}), \tag{3.135}$$

where

$$g_{10} = -\frac{m}{c}\left(\frac{\gamma}{\lambda^2} + \psi(l)\right), \quad g_{11}(\tilde{\xi}) = \frac{m}{c}\left(\frac{\gamma}{\lambda}\tilde{\phi} + \frac{\gamma}{\lambda^2}\tilde{\omega} + \lambda\tilde{\eta}(l) + \tilde{p}\right),$$
$$g_{20} = \frac{J\psi'(l)}{c}, \quad g_{21}(\tilde{\xi}) = -\frac{J}{c}(\lambda\tilde{\eta}'(l) + \tilde{q}).$$

Therefore, if $\eta \in H^4(0,l)$ is a solution of the boundary value problem (3.131), (3.132), (3.134), (3.135) for given $\tilde{\xi} \in X$, λ, and u, then formulas (3.129), (3.130), and (3.133) define the components $\phi, \omega, \zeta, p, q$ of a solution $\xi \in D(A)$ of Eq. (3.125). Vice versa, if $\xi \in D(A)$ is a solution of Eq. (3.125) then its component $\eta \in H^4(0,l)$ is a solution of the boundary value problem (3.131), (3.132), (3.134), (3.135) and the rest of its components ξ satisfy relations (3.129), (3.130), and (3.133).

We compute the general solution of Eq. (3.131) with boundary conditions (3.132) by using the method of variation of constants:

$$\eta(x) = \frac{1}{4\mu^3} \int_0^x \left(f_0(s)u + f_1(s, \tilde{\xi}) \right) K(x - s)\, ds + C_1\eta_1(x) + C_2\eta_2(x). \quad (3.136)$$

Here we use an auxiliary parameter $\mu > 0$:

$$\mu = \left(\frac{\lambda^2 \rho}{4c} \right)^{1/4}, \quad (3.137)$$

$$K(x) = \sin \mu x \operatorname{ch} \mu x - \cos \mu x \operatorname{sh} \mu x,$$
$$\eta_1(x) = \sin \mu x \operatorname{sh} \mu x, \quad \eta_2(x) = \cos \mu x \operatorname{sh} \mu x - e^{\mu x} \sin \mu x.$$

Since the smooth function $K(x)$ satisfies conditions

$$K(0) = K'(0) = K''(0) = K'''(0) = 0,$$

we conclude that the jth derivative of function (3.136) with respect to x can be represented as follows:

$$\eta^{(j)}(x) = \frac{1}{4\mu^3} \int_0^x \left(f_0(s)u + f_1(s, \tilde{\xi}) \right) K^{(j)}(x - s)\, ds + C_1\eta_1^{(j)}(x) + C_2\eta_2^{(j)}(x),$$

$$(3.138)$$

for $j = \overline{0, 3}$. For any $\tilde{\xi} \in X$ and $\lambda > 0$, functions $f_0(\cdot)$ and $f_1(\cdot, \tilde{\xi})$ belong to $L^2(0, l)$. Therefore, representations (3.138) and Eq. (3.131) imply that the solution $\eta(x)$ defined by (3.136) belongs to $H^4(0, l)$ for any C_1, C_2, u.

To eliminate parameters C_1, C_2, u from the solution $\eta(x)$, we substitute expressions (3.138) into the boundary conditions (3.134), (3.135) and the feedback control (3.112). Then we get a system of three linear algebraic equations with respect to $C_1, C_2,$ and u:

$$W \begin{pmatrix} C_1 \\ C_2 \\ u \end{pmatrix} = \begin{pmatrix} v_1(\tilde{\xi}) \\ v_2(\tilde{\xi}) \\ v_3(\tilde{\xi}) \end{pmatrix}, \quad (3.139)$$

where the components of the matrix $W = (w_{ij})$ and of the right-hand side $v_i(\tilde{\xi})$ are defined as follows:

$$w_{11} = -2\,\mu^3 \left(\sin \mu l \operatorname{ch} \mu l - \cos \mu l \operatorname{sh} \mu l \right) - \frac{4\mu^4 m}{\rho} \sin \mu l \operatorname{sh} \mu l,$$

$$w_{12} = -2\,\mu^3 \left(e^{\mu l}(\cos \mu l - \sin \mu l) + \cos \mu l \operatorname{ch} \mu l + \sin \mu l \operatorname{sh} \mu l \right)$$

$$- \frac{4\mu^4 m}{\rho} \left(\cos\mu l \, \mathrm{sh}\mu l - e^{\mu l} \sin\mu l \right),$$

$$w_{13} = \int_0^l f_0(s) \cos\mu \, (s - l) \, \mathrm{ch}\mu \, (s - l) \, ds - g_{10}$$

$$- \frac{\mu m}{\rho} \int_0^l f_0(s) \, (\cos\mu \, (s - l) \, \mathrm{sh}\mu \, (s - l) - \sin\mu \, (s - l) \, \mathrm{ch}\mu \, (s - l)) \, ds,$$

$$w_{21} = 2\mu^2 \cos\mu l \, \mathrm{ch}\mu l + \frac{4\mu^5 J}{\rho} \, (\cos\mu l \, \mathrm{sh}\mu l + \sin\mu \, l \, \mathrm{ch}\mu l),$$

$$w_{22} = -2\,\mu^2 \left(\sin\mu l \, \mathrm{ch}\mu l + e^{\mu l} \cos\mu l \right)$$

$$- \frac{4\mu^5 J}{\rho} \left(e^{\mu l} (\sin\mu l + \cos\mu l) + \sin\mu l \, \mathrm{sh}\mu l - \cos\mu l \, \mathrm{ch}\mu l \right),$$

$$w_{23} = -\frac{1}{2\mu} \int_0^l f_0(s) \, (\cos\mu \, (s - l) \, \mathrm{sh}\mu \, (s - l) + \sin\mu \, (s - l) \, \mathrm{ch}\mu \, (s - l)) \, ds$$

$$+ \frac{2\mu^2 J}{\rho} \int_0^l f_0(s) \sin\mu \, (s - l) \, \mathrm{sh}\mu \, (s - l) \, ds - g_{20},$$

$$w_{31} = 2\mu^2 c \left(\int_0^l \psi''(x) \cos\mu x \, \mathrm{ch}\mu x \, dx + \psi'(0) \right) - \gamma \, m \sin\mu l \, \mathrm{sh}\mu l$$

$$+ \frac{\gamma \rho}{4\mu} \left(e^{\mu l} (\cos\mu l - \sin\mu l) - e^{-\mu l} (\cos\mu l + \sin\mu l) \right),$$

$$w_{32} = 4\mu^3 c \psi\,(0) - 2\mu^2 c \left(\int_0^l (\sin\mu x \, \mathrm{ch}\mu x + e^{\mu x} \cos\mu x) \, \psi''(x) \, dx + \psi'(0) \right)$$

$$+ \gamma \, m \left(e^{\mu l} \sin\mu l - \cos\mu l \, \mathrm{sh}\mu l \right)$$

$$- \frac{\gamma \rho}{4\mu} \left(e^{\mu l} (3\cos\mu l - \sin\mu l) + e^{-\mu l} (\cos\mu l - \sin\mu l) - 4 \right),$$

$$w_{33} = \beta + \frac{c}{2\mu} \int_0^l \int_0^x f_0(s) \, (\cos\mu \, (x - s) \, \mathrm{sh}\mu \, (x - s) + \sin\mu \, (x - s) \, \mathrm{ch}\mu \, (x - s)) \, \psi''(x) \, ds \, dx$$

$$+ \frac{k}{2\mu^2} \sqrt{\frac{\rho}{c}} + \frac{\gamma\rho}{4\mu^3} \int_0^l \int_0^x f_0(s) \, (\cos\mu \, (x - s) \, \mathrm{sh}\mu \, (x - s) - \sin\mu \, (x - s) \, \mathrm{ch}\mu \, (x - s)) \, ds \, dx$$

$$+ \frac{\gamma \, m}{4\mu^3} \int_0^l f_0(s) \, (\sin\mu \, (s - l) \, \mathrm{ch}\mu \, (s - l) - \cos\mu \, (s - l) \, \mathrm{sh}\mu \, (s - l)) \, ds + \frac{\alpha^* \rho}{\mu^4 c},$$

$$v_1(\tilde{\xi}) = - \int_0^l f_1(s, \tilde{\xi}) \cos\mu \, (s - l) \, \mathrm{ch}\mu \, (s - l) \, ds$$

$$+ \frac{\mu m}{\rho} \int_0^l f_1(s, \tilde{\xi}) \, (\cos\mu \, (s - l) \, \mathrm{sh}\mu \, (s - l) - \sin\mu \, (s - l) \, \mathrm{ch}\mu \, (s - l)) \, ds + g_{11}(\tilde{\xi}),$$

$$v_2(\tilde{\xi}) = \frac{1}{2\mu} \int_0^l f_1(s, \tilde{\xi}) \left(\cos \mu \, (s - l) \, \mathrm{sh}\mu \, (s - l) + \sin \mu \, (s - l) \, \mathrm{ch}\mu \, (s - l)\right) ds$$

$$- \frac{2\mu^2 J}{\rho} \int_0^l f_1(s, \tilde{\xi}) \sin \mu \, (s - l) \, \mathrm{sh}\mu \, (s - l) \, ds + g_{21}(\tilde{\xi}),$$

$$v_3(\tilde{\xi}) = -\frac{c}{2\mu} \int_0^l \int_0^x f_1(s, \tilde{\xi}) \left(\cos \mu \, (x - s) \, \mathrm{sh}\mu \, (x - s) + \sin \mu \, (x - s) \, \mathrm{ch}\mu \, (x - s)\right) \psi''(x) \, ds \, dx$$

$$+ \frac{\gamma \rho}{4\mu^3} \int_0^l \int_0^x f_1(s, \tilde{\xi}) \left(\sin \mu \, (x - s) \, \mathrm{ch}\mu \, (x - s) - \cos \mu \, (x - s) \, \mathrm{sh}\mu \, (x - s)\right) ds \, dx$$

$$+ \frac{\gamma m}{4\mu^3} \int_0^l f_1(s, \tilde{\xi}) \left(\cos \mu \, (s - l) \, \mathrm{sh}\mu \, (s - l) - \sin \mu \, (s - l) \, \mathrm{ch}\mu \, (s - l)\right) ds$$

$$+ \frac{\alpha^* \tilde{\phi} + k\tilde{\omega}}{2\mu^2} \sqrt{\frac{\rho}{c}} + \frac{\alpha^* \rho \, \tilde{\omega}}{4\mu^4 c}, \tag{3.140}$$

$$\alpha^* = \alpha - \gamma \left(\rho \int_0^l \psi(x) \, dx + m\psi(l)\right) > 0.$$

In the above formulas, the parameter λ is expressed in terms of μ by relation (3.137), and the values of $f_0(s)$, $f_1(s, \tilde{\xi})$, g_{10}, $g_{11}(\tilde{\xi})$, g_{20}, $g_{21}(\tilde{\xi})$ have been defined earlier.

For given constants $c > 0$, $m > 0$, $J > 0$, $\rho > 0$, γ, and the function $\psi \in C^2[0, l]$, components of the matrix W depend on the parameter μ. Let us expand the determinant of W with respect to μ:

$$\det(W) = \frac{2\rho \, \alpha^*}{c} \mu + o(\mu) \tag{3.141}$$

as $\mu \to 0$.

We choose $\mu > 0$ small enough in order to satisfy the condition $\det(W) \neq 0$ according to (3.141). In the sequel, we assume that the number $\lambda > 0$ is fixed and corresponds to μ by formula (3.137), and that the parameter u is associated with ξ by linear functional (3.112). Condition $\det(W) \neq 0$ implies that, for any $\tilde{\xi} \in X$, there is a unique element $\xi \in D(A)$ satisfying Eq. (3.125) with (3.112). It means that $\xi = (\tilde{A} - \lambda I)^{-1}\tilde{\xi}$ defines the resolvent $(\tilde{A} - \lambda I)^{-1} : X \to X$ of $\tilde{A}$ for chosen $\lambda > 0$. Let us write the resolvent as

$$(\tilde{A} - \lambda I)^{-1}\tilde{\xi} = \left(R_\eta(\tilde{\xi}), R_\zeta(\tilde{\xi}), R_\phi(\tilde{\xi}), R_\omega(\tilde{\xi}), R_p(\tilde{\xi}), R_q(\tilde{\xi})\right)^T, \tag{3.142}$$

where the linear operators

$$R_\eta : X \rightarrow \{\eta \in H^2(0,l) : \eta(0) = \eta'(0) = 0\},$$

$$R_\zeta : X \rightarrow \{\zeta \in L^2(0,l)\}$$

and the functionals

$$R_\phi : X \rightarrow \{\phi \in \mathbb{R}\}, \quad R_\omega : X \rightarrow \{\omega \in \mathbb{R}\},$$

$$R_p : X \rightarrow \{p \in \mathbb{R}\}, \quad R_q : X \rightarrow \{q \in \mathbb{R}\}$$

are defined by formulas (3.129), (3.130), (3.133), (3.136), and (3.139).

Under the above assumptions, solutions C_1, C_2, u of system (3.139) satisfy the following estimate, for any $v_1(\tilde{\xi})$, $v_2(\tilde{\xi})$, $v_2(\tilde{\xi})$:

$$|C_1| + |C_2| + |u| \le M_1(|v_1(\tilde{\xi})| + |v_2(\tilde{\xi})| + |v_3(\tilde{\xi})|), \tag{3.143}$$

where the positive constant M_1 is defined by the norm of the matrix W^{-1}. According to the Cauchy–Schwartz inequality and the Friedrichs–Wirtinger type inequalities (3.103)–(3.104), formula (3.140) implies that $v_j : X \rightarrow \mathbb{R}$ are bounded linear functionals ($j = 1, 2, 3$). Therefore, estimate (3.143) ensures the boundedness of the linear map $\tilde{\xi} \mapsto (C_1, C_2, u)$ defined by system (3.139):

$$|C_1| + |C_2| + |u| \le M_2 \|\tilde{\xi}\|_X, \quad \forall \tilde{\xi} \in X, \tag{3.144}$$

with some constant $M_2 \ge 0$. Therefore, the functionals $R_\phi : \tilde{\xi} \mapsto \phi$, $R_\omega : \tilde{\xi} \mapsto \omega$ defined in (3.129) are bounded linear functionals from X to $\mathbb{R}$. By using the Cauchy–Schwartz inequality (3.138), we get the estimate

$$\|\eta^{(j)}\|_{L^2(0,l)} \le \frac{1}{4\mu^3} \left(\int_0^l \left(\int_0^x \left(f_0(s)u + f_1(s,\tilde{\xi}) \right) K^{(j)}(x-s)\, ds \right)^2 dx \right)^{1/2}$$

$$+ |C_1| \cdot \|\eta_1^{(j)}\|_{L^2(0,l)} + |C_2| \cdot \|\eta_2^{(j)}\|_{L^2(0,l)}$$

$$\le \frac{\sqrt{l}}{4\mu^3} \|f_0(\cdot)u + f_1(\cdot, \tilde{\xi})\|_{L^2} + |C_1| \cdot \|\eta_1^{(j)}\|_{L^2} + |C_2| \cdot \|\eta_2^{(j)}\|_{L^2}.$$

Hence, estimate (3.144) implies the existence of constant $M_3 \ge 0$:

$$\|\eta\|_{H^3(0,l)} = \left(\sum_{j=0}^3 \|\eta^{(j)}\|_{L^2(0,l)}^2 \right)^{1/2} \le M_3 \|\tilde{\xi}\|_X. \tag{3.145}$$

By the embedding theorem (see [34]), the space $H^3(0, l)$ is compactly embedded into $H^2(0, l)$, i.e. the linear operator $R_\eta : \tilde{\xi} \mapsto \eta$ defined by (3.136), (3.139) maps bounded subsets of X into precompact subsets of $H^2(0, l)$ according to (3.145).

Formulas (3.130) and (3.145) imply the estimate

$$\|\zeta\|_{H^2(0,l)} \le \|\tilde{\eta}\|_{H^2(0,l)} + \lambda\|\eta\|_{H^2(0,l)} \le \|\tilde{\eta}\|_{H^2(0,l)} + \lambda M_3\|\tilde{\xi}\|_X \le M_4\|\tilde{\xi}\|_X$$

with some positive constant M_4. Therefore, the linear operator $R_\zeta : \tilde{\xi} \mapsto \zeta$ from X to $L^2(0, l)$ is compact due to formulas (3.130), (3.136), and (3.139).

Let us estimate components p and q of the vector ξ in (3.133) by using boundary conditions $\eta(0) = \eta'(0) = \tilde{\eta}(0) = \tilde{\eta}'(0) = 0$:

$$|p| \le |\tilde{\eta}(l)| + \lambda|\eta(l)| = \left|\int_0^l \tilde{\eta}'(x)dx\right| + \lambda\left|\int_0^l \eta'(x)dx\right|$$

$$\le \|1\|_{L^2(0,l)} \cdot \left(\|\tilde{\eta}'\|_{L^2(0,l)} + \lambda\|\eta'\|_{L^2(0,l)}\right),$$

$$|q| \le |\tilde{\eta}'(l)| + \lambda|\eta'(l)| \le \|1\|_{L^2(0,l)} \cdot \left(\|\tilde{\eta}''\|_{L^2(0,l)} + \lambda\|\eta''\|_{L^2(0,l)}\right).$$

The right-hand side of each of these inequalities is less than $M_5\|\tilde{\xi}\|_X$ for some constant $M_5 > 0$ due to estimate (3.145) and the Friedrichs–Wirtinger type inequalities (3.103)–(3.104). Such estimates prove the boundedness of linear functionals R_p and R_q.

Thus, in notations (3.142), R_η and R_ζ are compact operators, R_ϕ, R_ω, R_p, and R_q are bounded functionals, therefore, the operator $(\tilde{A} - \lambda I)^{-1} : X \to X$ is compact.

For the equivalent norm $\|\cdot\|_V$ in X, the following properties hold: the semigroup $\{e^{t\tilde{A}}\}_{t \ge 0}$ is contractive and the operator $\tilde{A}$ is dissipative ($-\tilde{A}$ is accretive). In addition, since $\overline{D(\tilde{A})} = X$, $\tilde{A}(0) = 0$, and $(\tilde{A} - \lambda I)^{-1}$ is a compact operator for chosen $\lambda > 0$, we conclude that each trajectory $\{\xi(t)\}_{t \ge 0} = \{e^{t\tilde{A}}\xi(0)\}_{t \ge 0}$ of differential equation (3.110) and (3.112) is precompact by the Dafermos–Slemrod theorem (Theorem 1.6).

We denote the ω-limit set of a trajectory $\{\xi(t)\}_{t \ge 0} = \{e^{t\tilde{A}}\xi_0\}_{t \ge 0}$ by $\Omega(\xi_0)$. As it has been shown earlier, the function $V(\xi(t)) \ge 0$ is non-increasing for $t \ge 0$ and $\dot{V}(\xi) = 0$ only for $\omega = 0$. By using the invariance principle (Lemma 2.2), we conclude with the following corollary of Theorem 3.4.

Corollary 3.1 Let α and β be sufficiently large constants, and let $k > 0$. Then, for each trajectory $\{\xi(t)\}_{t \ge 0}$ of differential equation (3.110) with the control $u = h(\xi)$ of form (3.112), its ω-limit set $\Omega(\xi_0) \ne \emptyset$ is a compact subset of

$$S_c = \{\xi \in X \,|\, \omega = 0, \ V(\xi) = c\}$$

for some constant $c \ge 0$. In addition, $\Omega(\xi_0)$ is invariant with respect to the semigroup $\{e^{t\tilde{A}}\}_{t \ge 0}$. Here the functional V is defined by formula (3.113).

The above result reduces the analysis of limit sets of the trajectories for the closed-loop system (3.110) and (3.112) to the study of invariant subsets of S_c. In particular, it is of interest for further research to verify the Barbashin–Krasovskii type conditions [37] (Theorem 2.4) in order to obtain sufficient conditions for strong or partial asymptotic stability of the solution $\xi = 0$. For this purpose we will study an auxiliary spectral problem in the next subsection.

3.5.6.1 Analysis of the Spectral Problem

Consider the Hilbert space

$$H = \left\{ \theta = \begin{pmatrix} \eta \\ y \\ z \end{pmatrix} : \eta \in L^2(0, l), \ y, z \in \mathbb{C} \right\}$$

with the inner product

$$\langle \theta_1, \theta_2 \rangle = \int_0^l \rho \eta_1 \bar{\eta}_2 \, dx + m y_1 \bar{y}_2 + J z_1 \bar{z}_2.$$

Assume that there is a linear unbounded operator $F : D(F) \to H$ such that

$$\theta = \begin{pmatrix} \eta \\ y \\ z \end{pmatrix} \mapsto F\theta = \begin{pmatrix} \frac{1}{\rho}(c\eta'')'' \\ -\frac{1}{m}(c\eta'')'(l) \\ \frac{c}{J}\eta''(l) \end{pmatrix}, \tag{3.146}$$

where the domain of definition of F is defined as

$$D(F) = \left\{ \theta : \eta \in H^4(0, l), \ \eta(0) = \eta'(0) = 0, \ \eta(l) = y, \ \eta'(l) = z \right\} \subset H.$$

Some important properties of the operator F are formalized in the following lemmas.

Lemma 3.2 *The operator $F : D(F) \to H$ is symmetric and positive.*

Proof First, we prove the symmetry property for F [38]. For this purpose, we should prove that for any elements

$$\theta_1 = \begin{pmatrix} \eta_1 \\ y_1 \\ z_1 \end{pmatrix} \in D(F), \quad \theta_2 = \begin{pmatrix} \eta_2 \\ y_2 \\ z_2 \end{pmatrix} \in D(F)$$

the following equality holds:

$$\langle F\theta_1, \theta_2 \rangle = \langle \theta_1, F\theta_2 \rangle.$$

Integrating by parts and applying the boundary conditions for $\eta_1(x)$ and $\eta_2(x)$ from $\theta_1, \theta_2 \in D(F)$, we get

$$\langle F\theta_1, \theta_2 \rangle = \int_0^l (c\eta_1'')'' \bar{\eta}_2 \, dx - (c\eta_1'')' \bar{\eta}_2 \Big|_{x=l} + c\eta_1'' \bar{\eta}_2' \Big|_{x=l}$$

$$= \left((c_1\eta_1'')' \bar{\eta}_2 - c\eta_1'' \bar{\eta}_2' \right) \Big|_{x=0}^l - (c\eta_1'')' \bar{\eta}_2 \Big|_{x=l} + c\eta_1'' \bar{\eta}_2' \Big|_{x=l} = \int_0^l c\eta_1'' \bar{\eta}_2'' \, dx.$$

$$(3.147)$$

Similarly,

$$\langle \theta_1, F\theta_2 \rangle = \int_0^l c\eta_1'' \bar{\eta}_2'' \, dx,$$

hence,

$$\langle F\theta_1, \theta_2 \rangle = \langle \theta_1, F\theta_2 \rangle, \quad \forall \theta_1, \theta_2 \in D(F).$$

By using representation (3.147) with $\theta_1 = \theta_2 = \theta \in D(F)$, we get

$$\langle F\theta, \theta \rangle = \int_0^l c \, |\eta''|^2 \, dx \geq 0,$$

therefore, F is positive.

In the sequel, we assume that $\rho = \text{const}, c = \text{const}$.

Lemma 3.3 $F^{-1} : H \to H$ *is a compact self-adjoint operator.*

Proof For an arbitrary element $\theta_2 \in H$, let us prove the solvability of the linear equation

$$F\theta_1 = \theta_2$$

with respect to $\theta_1 \in D(F)$. We rewrite this equation for the components of vectors θ_1 and θ_2:

$$\eta_1''''(x) = \frac{\rho}{c} \eta_2(x), \quad x \in (0, l), \tag{3.148}$$

$$\eta_1'''(l) = -\frac{m}{c}y_2, \tag{3.149}$$

$$\eta_1''(l) = \frac{J}{c}z_2, \tag{3.150}$$

We also take into account the following boundary conditions from $\theta_1 \in D(F)$:

$$\eta_1(0) = \eta_1'(0), \tag{3.151}$$

$$y_1 = \eta_1(l), \quad z_1 = \eta_1'(l). \tag{3.152}$$

We write the general solution of Eq. (3.148) with boundary conditions (3.151) by using the method of variation of parameters:

$$\eta_1(x) = \frac{a_2}{2}x^2 + \frac{a_3}{6}x^3 + \frac{\rho}{6c}\int_0^x (x-s)^3\eta_2(s)\,ds, \quad x \in [0, l], \tag{3.153}$$

where a_2, a_3 are arbitrary constants. For any $\eta_2 \in L^2(0, l)$, the formula (3.153) defines the function $\eta_1 \in H^4(0, l)$ because the derivatives of η_1 can be represented as

$$\eta_1'(x) = a_2 x + \frac{a_3}{2}x^2 + \frac{\rho}{2c}\int_0^x (x-s)^2\eta_2(s)\,ds,$$

$$\eta_1''(x) = a_2 + a_3 x + \frac{\rho}{c}\int_0^x (x-s)\eta_2(s)\,ds,$$

$$\eta_1'''(x) = a_3 + \frac{\rho}{c}\int_0^x \eta_2(s)\,ds.$$

We define constants a_2 and a_3 from boundary conditions (3.149) and (3.150):

$$a_3 = -\frac{\rho}{c}\int_0^l \eta_2(s)\,ds - \frac{m}{c}y_2, \tag{3.154}$$

$$a_2 = -la_3 - \frac{\rho}{c}\int_0^l (l-s)\eta_2(s)\,ds + \frac{J}{c}z_2. \tag{3.155}$$

Thus, each element $\theta_2 \in H$ is related to a_2 and a_3 by formulas (3.154) and (3.155) and to y_1 and z_1 by formulas (3.152). Numbers a_2 and a_3 define the function $\eta_1 \in H^4(0, l)$ by formula (3.153). The above relations define the operator

$$F^{-1} : \theta_2 \in H \mapsto \theta_1 \in D(F) \subset H.$$

The Riesz theorem implies that the linear functionals $\theta_2 \mapsto a_2$ and $\theta_2 \mapsto a_3$ defined by formulas (3.154) and (3.155) are bounded. By using representation (3.153) we prove that the components of F^{-1} are bounded: the map $\theta_2 \in H \mapsto \eta_1 \in H^4(0, l)$ is bounded, and the functionals $\theta_2 \in H \mapsto y_1, \theta_2 \in H \mapsto y_1$ are bounded in (3.152). Since the Sobolev space $H^4(0, l)$ is compactly embedded into $L^2(0, l)$ (cf. [34]), such arguments prove the compactness of the operator $F^{-1} : H \to H$. The symmetry of F yield that $F^{-1} : H \to H$ is a self-adjoint operator [38].

By the theorem on the spectrum of a compact self-adjoint operator (see [38]), we conclude that there is an orthonormal basis of the Hilbert space H:

$$\theta_1, \theta_2, \ldots, \theta_n, \ldots,$$

which consists of the eigenvectors of F^{-1} corresponding to its eigenvalues

$$\lambda_1 \geq \lambda_2 \geq \cdots \geq \lambda_n \geq \cdots .$$

As the operator F^{-1} is invertible, $\lambda = 0$ is not an eigenvalue of F^{-1}. Therefore, all λ_n are positive because F is positive. The inverse operator definition implies that θ_n is an eigenvalue of the operator F:

$$F\theta_n = \mu_n\theta_n, \quad \theta_n \in D(F), \tag{3.156}$$

where $\mu_n = 1/\lambda_n > 0$. Let us rewrite (3.156) by using the notation

$$\theta_n = \begin{pmatrix} \eta \\ \eta(l) \\ \eta'(l) \end{pmatrix} \in D(F),$$

where $\eta \in H^4(0, l)$. Then we get the following spectral problem with a parameter $\mu = \mu_n$:

$$\eta''''(x) = \frac{\mu\rho}{c}\eta(x), \quad x \in (0, l), \tag{3.157}$$

$$\eta(0) = \eta'(0) = 0, \tag{3.158}$$

$$\eta'''(l) = -\frac{\mu m}{c}\eta(l), \tag{3.159}$$

$$\eta''(l) = \frac{\mu J}{c}\eta'(l). \tag{3.160}$$

Since all eigenvalues μ_n of the operator F are real and positive, we will consider problem (3.157)–(3.160) with real functions $\eta(x)$ only.

Lemma 3.4 *For each eigenvalue $\mu = \mu_n > 0$ of the spectral problem (3.157) and (3.160), there is a unique (up to a constant multiplier) eigenfunction $\eta(x)$, $0 \le x \le l$. Moreover, $\eta''(0) \neq 0$.*

Proof For an arbitrary $\mu > 0$, we denote

$$\tilde{\beta} = l\left(\frac{\mu\rho}{c}\right)^{1/4} > 0. \tag{3.161}$$

Then the general solution of differential equation (3.157) has the following form

$$\eta(x) = C_1 \cos\left(\frac{\tilde{\beta}x}{l}\right) + C_2 \sin\left(\frac{\tilde{\beta}x}{l}\right) + C_3 \mathrm{ch}\left(\frac{\tilde{\beta}x}{l}\right) + C_4 \mathrm{sh}\left(\frac{\tilde{\beta}x}{l}\right),$$

where C_1, C_2, C_3, and C_4 are arbitrary constants. Substituting this solution into (3.158), we get $C_3 = -C_1$ and $C_4 = -C_2$. Then any solution of Eq. (3.157) with boundary conditions (3.158) can be represented as

$$\eta(x) = C_1\left\{\cos\left(\frac{\tilde{\beta}x}{l}\right) - \mathrm{ch}\left(\frac{\tilde{\beta}x}{l}\right)\right\} + C_2\left\{\sin\left(\frac{\tilde{\beta}x}{l}\right) - \mathrm{sh}\left(\frac{\tilde{\beta}x}{l}\right)\right\}. \tag{3.162}$$

By differentiating the function $\eta(x)$, we get

$$\frac{l\eta'(x)}{\tilde{\beta}} = -C_1\left\{\sin\left(\frac{\tilde{\beta}x}{l}\right) + \mathrm{sh}\left(\frac{\tilde{\beta}x}{l}\right)\right\} + C_2\left\{\cos\left(\frac{\tilde{\beta}x}{l}\right) - \mathrm{ch}\left(\frac{\tilde{\beta}x}{l}\right)\right\},$$

$$\frac{l^2\eta''(x)}{\tilde{\beta}^2} = -C_1\left\{\cos\left(\frac{\tilde{\beta}x}{l}\right) + \mathrm{ch}\left(\frac{\tilde{\beta}x}{l}\right)\right\} - C_2\left\{\sin\left(\frac{\tilde{\beta}x}{l}\right) + \mathrm{sh}\left(\frac{\tilde{\beta}x}{l}\right)\right\},$$

$$\frac{l^3\eta'''(x)}{\tilde{\beta}^3} = C_1\left\{\sin\left(\frac{\tilde{\beta}x}{l}\right) - \mathrm{sh}\left(\frac{\tilde{\beta}x}{l}\right)\right\} - C_2\left\{\cos\left(\frac{\tilde{\beta}x}{l}\right) + \mathrm{ch}\left(\frac{\tilde{\beta}x}{l}\right)\right\}.$$

$$\tag{3.163}$$

Then boundary conditions (3.159) and (3.160) take the form

$$C_1\left\{\frac{m\tilde{\beta}}{\rho l}(\cos\tilde{\beta} - \mathrm{ch}\tilde{\beta}) + \sin\tilde{\beta} - \mathrm{sh}\tilde{\beta}\right\} + C_2\left\{\frac{m\tilde{\beta}}{\rho l}(\sin\tilde{\beta} - \mathrm{sh}\tilde{\beta}) - \cos\tilde{\beta} - \mathrm{ch}\tilde{\beta}\right\} = 0$$

$$\tag{3.164}$$

and

$$
C_1 \left\{ -\frac{J\tilde{\beta}^3}{\rho l^3}(\sin\tilde{\beta} + \mathrm{sh}\tilde{\beta}) + \cos\tilde{\beta} + \mathrm{ch}\tilde{\beta} \right\} + C_2 \left\{ \frac{J\tilde{\beta}^3}{\rho l^3}(\cos\tilde{\beta} - \mathrm{ch}\tilde{\beta}) + \sin\tilde{\beta} + \mathrm{sh}\tilde{\beta} \right\} = 0,
$$
(3.165)

respectively. Thus, we get the system of two linear homogeneous equations (3.164), (3.165) with respect to C_1, C_2. Then we compute the determinant Δ of the matrix of this system and equate it to zero:

$$
\frac{\Delta}{2} = \tilde{m}\tilde{J}\tilde{\beta}^4(1 - \mathrm{ch}\tilde{\beta}\cos\tilde{\beta}) - \tilde{J}\tilde{\beta}^3(\mathrm{sh}\tilde{\beta}\cos\tilde{\beta} + \mathrm{ch}\tilde{\beta}\sin\tilde{\beta})
$$
$$
+ \tilde{m}\tilde{\beta}(\mathrm{sh}\tilde{\beta}\cos\tilde{\beta} - \mathrm{ch}\tilde{\beta}\sin\tilde{\beta}) + \mathrm{ch}\tilde{\beta}\cos\tilde{\beta} + 1 = 0,
$$
(3.166)

where

$$
\tilde{m} = \frac{m}{\rho l}, \quad \tilde{J} = \frac{J}{\rho l^3}.
$$

Nontrivial solutions $\eta(x)$ of the problem (3.157)–(3.160) exist only for $\tilde{\beta}$ satisfying the characteristic equation (3.166). If $\mu > 0$ is an eigenvalue of the problem (3.157)–(3.160) then the corresponding value of $\tilde{\beta}$ is a root of Eq. (3.166) by formula (3.161). Besides, each eigenfunction $\eta(x)$ corresponding to μ can be represented in the form (3.162), where C_1 and C_2 satisfy the algebraic system (3.164), (3.165). Let us note that the coefficient at C_2 in Eq. (3.164) is non-zero for any $\tilde{\beta} > 0$. Indeed, that coefficient is equal to zero only if

$$
\frac{\sin\tilde{\beta} - \mathrm{sh}\tilde{\beta}}{\cos\tilde{\beta} + \mathrm{ch}\tilde{\beta}} = \frac{\rho l}{m\tilde{\beta}}.
$$
(3.167)

If $\tilde{\beta} > 0$ then $\mathrm{sh}\tilde{\beta} > \sin\tilde{\beta}$ and the left-hand side of relation (3.167) is negative, which is impossible due to the positivity of its right-hand side. Thus, Eq. (3.164) can be uniquely solved with respect to C_2 for any $\tilde{\beta} > 0$:

$$
C_2 = C_1 \frac{m\tilde{\beta}(\cos\tilde{\beta} - \mathrm{ch}\tilde{\beta}) + \rho l(\sin\tilde{\beta} - \mathrm{sh}\tilde{\beta})}{m\tilde{\beta}(\mathrm{sh}\tilde{\beta} - \sin\tilde{\beta}) + \rho l(\cos\tilde{\beta} + \mathrm{ch}\tilde{\beta})}.
$$
(3.168)

We see that expressions (3.162), (3.168) define the eigenfunction $\eta(x)$ of the spectral problem up to the constant $C_1 \neq 0$ (if $\tilde{\beta}$ is not a root of the characteristic equation). Hence, each eigenvalue $\mu > 0$ corresponds to a one-dimensional invariant subspace for the problem (3.157)–(3.160).

It remains to prove that $\eta''(0) \neq 0$. If we assume that $\eta''(0) = 0$ for an eigenfunction $\eta(x)$, then formula (3.163) yields

$$C_1 = -\frac{l^2 \eta''(0)}{2\tilde{\beta}^2} = 0.$$

Then it follows from relation (3.168) that $C_2 = 0$, therefore, $\eta(x) \equiv 0$. This contradicts the definition of an eigenfunction. So we conclude that $\eta''(0) = 0$ for each eigenfunction of the spectral problem (3.157)–(3.160).

3.5.6.2 Asymptotic Stability Proof

In Theorem 3.3, we have constructed the positive definite Lyapunov functional $V : X \to \mathbb{R}$ such that its time derivative along the trajectories of the closed-loop system (3.124) takes the form

$$\dot{V}(\xi) = -k\omega^2 \le 0. \tag{3.169}$$

All trajectories $\{e^{t\tilde{A}} \xi_0 \mid t \ge 0\}$ are precompact in X by Theorem 3.4. Therefore, in order to prove the asymptotic stability, it suffices to verify that $\{0\}$ is the unique subset of

$$Z_0 = \{\xi \mid \dot{V}(\xi) = 0\},$$

which is invariant with respect to the semigroup $\{e^{t\tilde{A}}\}_{t \ge 0}$. Then the solution $\xi = 0$ of the closed-loop system (3.124) is asymptotically stable with respect to the functional $\mu(\xi) = \|\xi\|$ (or strongly asymptotically stable) by Theorem 2.4.

Let $\xi(t)$ be a solution of Eq. (3.124) and $\xi(t) \in Z_0$ for all $t \ge 0$. Then formula (3.169) implies $\omega(t) \equiv 0$ for the corresponding component of $\xi(t)$, so $\phi(t) \equiv \text{const}$. Moreover, equation (3.112) yields $u = \frac{d\omega(t)}{dt} \equiv 0$, therefore, $h(\xi(t)) = 0$. With $\xi \in X$, we associate the element $\theta \in H$ by the formula

$$\theta = \begin{pmatrix} \eta \\ \eta(l) \\ \eta'(l) \end{pmatrix}.$$

If $\xi(t)$ is a solution of Eq. (3.110) with $\phi(t) = \text{const}$, $u = 0$, then it is easy to show that the corresponding function $\theta(t)$ satisfies the following differential equation in H:

$$\frac{d^2}{dt^2}\theta(t) = -F\theta(t) + b\phi, \tag{3.170}$$

where the operator $F : D(F) \to H$ is defined by formula (3.146) and $b = (\gamma, \gamma, 0)^T \in H$.

Let $\theta(t)$ be a solution of differential equation (3.170) for $t \ge 0$. As the eigenvectors $\{\theta_j\}$ of F define an orthonormal basis in H then $\theta(t)$ can be uniquely expanded into the Fourier series

$$\theta(t) = \sum_{j=1}^{\infty} q_j(t)\theta_j, \tag{3.171}$$

where the Fourier coefficient $q_j(t)$ plays a role of the jth modal coordinate,

$$\theta_j = \begin{pmatrix} \eta_j \\ \eta_j(l) \\ \eta_j'(l) \end{pmatrix}, \quad \|\theta_j\|_H = 1,$$

and $\eta_j \in H^4(0, l)$ is an eigenfunction of the spectral problem (3.157)–(3.160). By multiplying both sides of Eq. (3.170) with θ_n and substituting $\theta(t)$ from formula (3.171), we get the following system of differential equations with respect to $\theta_n(t)$:

$$\frac{d^2}{dt^2} q_n(t) = -\mu_n q_n(t) + \phi b_n, \quad n = 1, 2, \ldots, \tag{3.172}$$

where $b_n = \langle b, \theta_n \rangle$, $\sum_{n=1}^{\infty} b_n^2 = \|b\|_H^2 < \infty$ due to Parseval's identity. By solving these equations, we obtain

$$q_n(t) = A_n \cos(\sqrt{\mu_n} t) + B_n \sin(\sqrt{\mu_n} t) + \frac{\phi b_n}{\mu_n} \tag{3.173}$$

with some constants A_n and B_n. If the function $\xi(t)$ corresponding to a solution $\theta(t)$ satisfies the condition $h(\xi(t)) \equiv 0$ for all $t \geq 0$, then representation (3.173) implies

$$\left(\alpha - \gamma \left(\int_0^l \rho \psi \, dx + m\psi(l) \right) \right) \phi$$

$$+ \sum_{n=1}^{\infty} q_n(t) \left\{ \int_0^l c\eta_n'' \psi'' dx + \left(c\eta_n'' \psi' - (c\eta_n'')' \psi \right) \Big|_{x=0} - \gamma \left(\int_0^l \rho \eta_n \, dx + m\eta_n(l) \right) \right\} \equiv 0.$$

For $\psi(x) = -x$, this expression reads as

$$\left(\alpha - \gamma \left(\int_0^l \rho \psi \, dx + m\psi(l) \right) \right) \phi - \sum_{n=1}^{\infty} q_n(t) \left\{ c\eta_n''(0) + \gamma \left(\int_0^l \rho \eta_n \, dx + m\eta_n(l) \right) \right\} \equiv 0. \tag{3.174}$$

It follows from representation (3.173) that the left-hand side of identity (3.174) is a linear combination of functions

$$1, \cos(\sqrt{\mu_1} t), \sin(\sqrt{\mu_1} t), \cos(\sqrt{\mu_2} t), \sin(\sqrt{\mu_2} t), \ldots \tag{3.175}$$

Lemma 3.4 implies that there are no multiple eigenvalues among μ_j, therefore, functions (3.175) are linearly independently on $t \in [0, +\infty)$ (see, e.g., [12, Theorem 1.2.17]). It means that all coefficients $A_n = B_n$ vanish and $\phi = 0$ in (3.173) (for $\alpha > 0$ large enough). So we conclude that $\theta(t) \equiv 0$ and $\phi(t) \equiv 0$, i.e. the maximal invariant subset of Z_0 is the singleton $\xi = 0$.

The above arguments prove that system (3.110) is strongly stabilizable in the following sense.

Theorem 3.5 *Suppose that α, β are sufficiently large constants and $\psi(x) = -x$. Then the solution $\xi = 0$ of the closed-loop system (3.124) is strongly asymptotically stable in the sense of Lyapunov.*

References

1. Lagnese, J.E., Leugering, G., Schmidt, E.J.P.G.: Modeling, Analysis and Control of Dynamic Elastic Multi-Link Structures. Springer, New York (1994)
2. Baillieul, J., Levi, M.: Rotational elastic dynamics. Physica D **27**, 43–62 (1987)
3. Baillieul, J., Levi, M.: Constrained relative motions in rotational mechanics. Arch. Ration. Mech. Anal. **115**, 101–135 (1991)
4. Bloch, A.M., Titi, E.S.: On the dynamics of rotating elastic beams. New Trends in Systems Theory, pp. 128–135. Birkhauser, Boston (1991). (Proc. Jt. Conf., Genoa, Italy, 1990)
5. Cai, G.-P., Hong, J.-Z., Yang, S.X.: Model study and active control of a rotating flexible cantilever beam. Int. J. Mech. Sci. **46**, 871–889 (2004)
6. Cai, G.-P., Hong, J.-Z., Yang, S.X.: Dynamic analysis of a flexible hub-beam system with tip mass. Mech. Res. Commun. **32**, 173–190 (2005)
7. Cai, G.-P., Lim, C.W.: Active control of a flexible hub-beam system using optimal tracking control method. Int. J. Mech. Sci. **48**, 1150–1162 (2006)
8. Cai, G.-P., Lim, C.W.: Dynamics studies of a flexible hub-beam system with significant damping effect. J. Sound Vib. **318**, 1–17 (2008)
9. Laousy, H., Xu, C.Z., Sallet, G.: Boundary feedback stabilization of a rotating body-beam system. IEEE Trans. Autom. Control **41**, 241–245 (1996)
10. Xu, C.Z., Baillieul, J.: Stabilizability and stabilization of a rotating body-beam system with torque control. IEEE Trans. Autom. Control **38**, 1754–1765 (1993)
11. Coron, J.-M., d'Andrea-Novel, B.: Stabilization of a rotating body beam without damping. IEEE Trans. Autom. Control **44**, 608–618 (1998)
12. Krabs, W.: On Moment Theory and Controllability of One Dimensional Vibrating Systems and Heating Processes. Lecture Notes in Control and Information Sciences, vol. 173 Springer, Berlin (1992)
13. Guo, B.-Z., Luo, Z.-H.: On the exponential stability of an initial-boundary equation arising from strain feedback control of flexible robot arms with rigid offset. Int. J. Control **69**, 227–238 (1998)
14. Halim, D., Moheimani, S.O.R.: Spatial H_2 control of a piezoelectric laminate beam: experimental implementation. IEEE Trans. Control Syst. Technol. **10**, 533–546 (2002)
15. Jarrar, F.S.M., Hamdan, M.N.: Nonlinear vibrations and buckling of a flexible rotating beam: a prescribed torque approach. Mech. Mach. Theor. **42**, 919–939 (2007)
16. Lee, Y.-S., Elliott, S.J.: Active position control of a flexible smart beam using internal model control. J. Sound Vib. **242**(5), 767–791 (2001)
17. Tavasoli, A., Eghtesad, M., Jafarian, H.: Two-time scale control and observer design for trajectory tracking of two cooperating robot manipulators moving a flexible beam. Robot. Auton. Syst. **57**, 212–221 (2009)

18. Yang, W., Zhang, Z., Shen, R.: Modeling of system dynamics of a slewing flexible beam with moving payload pendulum. Mech. Res. Commun. **34**, 260–266 (2007)
19. Junkins, J.L., Kim, Y.: Introduction to Dynamics and Control of Flexible Structures. AIAA Education Series. AIAA, Reston (1993)
20. Nabiullin, M.K.: Stationary Motions and Stability of Elastic Satellites (in Russian). Nauka, Novosibirsk (1990)
21. Zuyev, A.: Partial asymptotic stabilization of nonlinear distributed parameter systems. Automatica **41**, 1–10 (2005)
22. Timoshenko, S.P.: On the correction for shear of the differential equation for transverse vibrations of prismatic bars. Phil. Mag. **XLI**, 744–746 (1921). Reprinted. In: The Collected Papers of Stephen P. Timoshenko. McGraw-Hill, London (1953)
23. Berdichevsky, V.: Variational Principles of Continuum Mechanics: I. Fundamentals. Interaction of Mechanics and Mathematics. Springer, Berlin (2010)
24. Goldstein, H.: Classical Mechanics, 2nd edn. Addison-Wesley, Massachusets (1980)
25. Luo, Z.-H., Guo, B.-Z., Morgul, O.: Stability and Stabilization of Infinite Dimensional Systems with Applications. Springer, London (1999)
26. Levan, N., Rigby, L.: Strong stabilizability of linear contractive control systems on Hilbert space. SIAM J. Control Optim. **17**, 23–35 (1979)
27. Curtain, R.F.: On stabilizability of linear spectral systems via state boundary feedback. SIAM J. Control Optim. **23**, 144–152 (1985)
28. Curtain, R.F., Zwart, H.: An Introduction to Infinite-Dimensional Linear Systems Theory. Springer, New York (1995)
29. Kreĭn, S.G.: Linear Differential Equations in Banach Space. AMS, Providence (1971)
30. Kantorovich, L.V., Akilov, G.P.: Functional Analysis. Pergamon Press, Oxford (1982)
31. Pazy, A.: Semigroups of Linear Operators and Applications to Partial Differential Equations. Springer, New York (1983)
32. King, B.B.: Modeling and control of a multiple component structure. J. Math. Syst. Estim. Control **4**(4), 1–36 (1994)
33. Zuyev, A.: Feedback stabilization of a system of rigid bodies with a flexible beam. In: Robot Motion and Control 2009. Lecture Notes in Control and Information Sciences, pp. 69–81. Springer, Berlin (2009)
34. Mikhajlov, V.P.: Partial Differential Equations (trans: Russian by P.C. Sinha). Revised from the 1976 Russian ed. Mir Publishers, Moscow (1978)
35. Hardy, G., Littlewood, J.E., Pólya, G.: Inequalities, 2nd edn. Cambridge University Press, Cambridge (1952)
36. Krabs, W., Sklyar, G.M.: On the controllability of a slowly rotating timoshenko beam. J. Anal. Appl. **18**, 437–448 (1999)
37. Krasovskii, N.N.: Problems of the Theory of Stability of Motion. Stanford University Press, Stanford (1963)
38. Akhiezer, N.I., Glazman, I.M.: Theory of Linear Operators in Hilbert Space. Dover Publications, New York (1993)

Chapter 4
Reachable Sets and Controllability Conditions

Abstract The approximate steering problem is considered in this chapter for a linear distributed parameter system with finite-dimensional control. An approach for solving this problem is proposed by using exact solutions of the steering problem for reduced systems and the spillover analysis. This approach allows also to estimate the reachable sets and to study the approximate controllability. To satisfy the spillover condition, we exploit L^2-optimal controls for a family of finite-dimensional subsystems. These controls are constructed explicitly for a system of oscillators with one-dimensional input. As a result, we obtain sufficient conditions for the approximate controllability in terms of the distribution of eigenfrequencies in such system. These conditions are applied for the approximate controllability study of a rotating body-beam system.

The problems of spectral, approximate, exact, and null controllability of distributed parameter systems have been intensively studied in the last few decades [1–5]. On the one hand, the question of the approximate controllability of a linear time-invariant system on a Hilbert space can be formulated in terms of an invariant subspace of the corresponding adjoint semigroup [2, 6, 7]. On the other hand, the problem of an effective control design remains challenging for a wide range of mechanical systems with distributed parameters. In the monograph [8, Chap. 7], the problem of steering the Euler–Bernoulli beam to its equilibrium state is studied by using the separation of variables method and the eigenfunction expansion. In order to solve this problem, an optimal control for the system with elastic coordinates has been obtained. Problems of null-controllability and spectral controllability of a network of the Euler–Bernoulli beams are addressed in the monograph [9]. The subspace of controllable initial states in some time $T > 0$ is defined for a network of beams controlled from one exterior node [9, Chap. 8]. The controllability problem of diagonal linear systems without zero eigenvalues has been considered in the paper [10].

The goal of this chapter is to propose a constructive control strategy, based on a reduced model, and to justify that this approach can be used to solve the steering problem and to check approximate controllability conditions for a class of infinite-dimensional systems.

© Springer International Publishing Switzerland 2015

A.L. Zuyev, *Partial Stabilization and Control of Distributed Parameter Systems with Elastic Elements*, Lecture Notes in Control and Information Sciences 458, DOI 10.1007/978-3-319-11532-0_4

4.1 Problem Statement and Preliminaries

Consider a linear control system in a Hilbert space H given by

$$\dot{x} = Ax + Bu, \quad x \in H, \ u \in \mathbb{R}^m. \tag{4.1}$$

Here $A : D(A) \to H$ is the infinitesimal generator of a C_0 semigroup of bounded linear operators $\{e^{tA}\}_{t \geq 0}$ on H, and $B : \mathbb{R}^m \to H$ is a bounded operator.

For $x^0 \in H$ and $u \in L^\infty(0, \tau)$, the mild solution $x(t; x^0, u)$ of (4.1) satisfying the initial condition $x|_{t=0} = x^0$ is

$$x(t; x^0, u) = e^{tA} x^0 + \int_0^t e^{(t-s)A} Bu(s) \, ds, \quad 0 \leq t \leq \tau. \tag{4.2}$$

Definition 4.1 For $x^0 \in H$, *reachable sets* of system (4.1) are defined as follows:

$$\mathcal{R}_\tau(x^0) = \{x(\tau; x^0, u) \mid u \in L^2(0, \tau)\} \quad \text{(reachable set from } x^0 \text{ at time } \tau \geq 0);$$

$$\mathcal{R}(x^0) = \bigcup_{\tau \geq 0} \mathcal{R}_\tau(x^0) \quad \text{(reachable set from } x^0).$$

Let us recall several notions of controllability for distributed parameter systems (see, e.g., [2, 11]).

Definition 4.2 System (4.1) is called

- *exactly controllable* if $\mathcal{R}(x^0) = H$ for each $x^0 \in H$;
- *approximately controllable* if $\overline{\mathcal{R}(x^0)} = H$ for each $x^0 \in H$;
- *exactly controllable in time τ* if $\mathcal{R}_\tau(x^0) = H$ for each $x^0 \in H$;
- *approximately controllable in time τ* if $\overline{\mathcal{R}_\tau(x^0)} = H$ for each $x^0 \in H$.

The goal of this section is to estimate the reachable sets of the infinite-dimensional system (4.1) by considering its finite-dimensional projections. To implement this idea, we will estimate the spillover effect for a family of controls that solve the steering problem for such projections. Note that the spillover analysis has been carried out for the stabilization problem in the paper [12].

The basic result of this section is as follows.

Theorem 4.1 *Let* $\{Q_N\}_{N=1}^\infty$ *be a family of bounded linear operators on H such that* e^{tA} *and* Q_N *commute, and let*

$$\lim_{N \to \infty} \|Q_N x\| = 0 \quad \text{for all } x \in H. \tag{4.3}$$

Assume that, for $x^0, x^1 \in H$ and any $N \geq 1$, there is a control $u^N \in L^2(0, \tau)$ such that

$$P_N\left(x(\tau; x^0, u^N) - x^1\right) = 0, \quad P_N = I - Q_N, \qquad (4.4)$$

and

$$\lim_{N \to \infty} \left(\|Q_N B\| \cdot \|u^N\|_{L^2(0,\tau)}\right) = 0. \qquad (4.5)$$

Then, for any $\varepsilon > 0$, there exists an $N_0(\varepsilon)$ such that

$$\|x(\tau; x^0, u^N) - x^1\| < \varepsilon \quad \text{for each } N \geq N_0(\varepsilon).$$

Proof For a given pair $x^0, x^1 \in H$ and a sequence $\{Q_N\}$, consider a family of controls $u^N = u^N_{x^0, x^1}(t)$ satisfying conditions (4.4) and (4.5). Then representation (4.2) yields

$$\|x(\tau; x^0, u^N) - x^1\| = \|Q_N x(\tau; x^0, u^N) - Q_N x^1\| \leq \|Q_N e^{\tau A} x^0\|$$

$$+ \|Q_N x^1\| + \left\| \int_0^\tau Q_N e^{(\tau-s)A} B u^N(s) ds \right\|.$$

By using the assumption that operators e^{tA} and Q_N commute and exploiting the Cauchy–Schwartz inequality, we get

$$\|x(\tau; x^0, u^N) - x^1\| \leq \|e^{\tau A}\| \cdot \|Q_N x^0\| + \|Q_N x^1\| + \sup_{t \in [0,\tau]} \|e^{tA}\| \cdot \|Q_N B\| \cdot \int_0^\tau |u^N(s)| ds$$

$$\leq \|Q_N x^1\| + \|e^{\tau A}\| \cdot \|Q_N x^0\| + \sqrt{\tau} \|Q_N B\| \sup_{t \in [0,\tau]} \|e^{tA}\| \cdot \|u^N\|_{L^2(0,\tau)}.$$

$$(4.6)$$

From Theorem 2.2 of [13, p. 4] it follows that there exist constants $\omega \geq 0$ and $M \geq 1$ such that

$$\|e^{tA}\| \leq M e^{\omega t} \quad \text{for} \quad 0 \leq t < \infty.$$

By using this estimate together with conditions (4.3) and (4.5), we conclude that, for any $\varepsilon > 0$, there is a number $N_0(\varepsilon) \geq 1$ such that

$$\|Q_N x^0\| < \frac{\varepsilon e^{-\omega \tau}}{3M}, \quad \|Q_N x^1\| < \frac{\varepsilon}{3},$$

$$\|Q_N B\| \cdot \|u^N\|_{L^2(0,\tau)} < \frac{\varepsilon e^{-\omega \tau}}{3M \sqrt{\tau}} \quad \text{for all } N \geq N_0(\varepsilon). \qquad (4.7)$$

Inequalities (4.6) and (4.7) show that

$$\|x(\tau; x^0, u^N) - x^1\| < \varepsilon \quad \text{for all } N \geq N_0(\varepsilon).$$

This completes the proof.

As it follows from the proof, inequalities (4.7) may be used for computing the required number N depending on x^0, x^1, and ε.

Theorem 4.1 implies the following corollaries.

Corollary 4.1 *If the assumptions of Theorem 4.1 are satisfied for each $x^0, x^1 \in X$, where X is a dense subset of H, then system (4.1) is approximately controllable in time τ.*

Corollary 4.2 *Assume that $x^0, x^1 \in H$, the assumptions of Theorem 4.1 are satisfied, and the family of controls $\{u^N(t)\}_{N=1}^{\infty}$ is precompact in $L^2(0, \tau)$. Then $x^1 \in \mathcal{R}_\tau(x^0)$.*

For a possible application of Theorem 4.1 and its corollaries, we assume that each operator $P_N = I - Q_N$ is a finite-dimensional projection. Let $\dim(\operatorname{Im} P_N) = d_N$. For given $x^0, x^1 \in H$, we introduce vectors

$$\tilde{x}_N^0 = P_N x^0, \quad \tilde{x}_N^1 = P_N x^1, \quad \tilde{x}_N = P_N x, \quad (\tilde{x}_N^0, \tilde{x}_N^1, \tilde{x}_N \in \operatorname{Im} P_N),$$

and operators

$$\tilde{A}_N = P_N A P_N, \quad \tilde{B}_N = P_N B.$$

Then condition (4.4) implies that $u_{x^0, x^1}^N(t)$ should solve the following control problem:

$$\dot{\tilde{x}}_N = \tilde{A}_N \tilde{x}_N + \tilde{B}_N u, \quad t \in [0, \tau], \tag{4.8}$$
$$\tilde{x}_N|_{t=0} = \tilde{x}_N^0, \quad \tilde{x}_N|_{t=\tau} = \tilde{x}_N^1.$$

Here we have used the assumption that P_N and A commute as well as the property $P_N = P_N^2$ of a projection. To satisfy condition (4.5), it is natural to look for an optimal control $u = u_{x^0, x^1}^N(t)$ that minimizes the functional

$$J = \int_0^\tau (Qu, u)\, dt \to \min \tag{4.9}$$

with some symmetric positive definite $m \times m$-matrix Q. As control system (4.8) evolves on a real d_N-dimensional vector space $\operatorname{Im} P_N$, we may treat (4.8) as a system on $\mathbb{R}^{d_N}$ without lack of generality. By applying the Pontryagin maximum principle, we get the optimal control $u = u^N(t)$ for problem (4.8) and (4.9):

$$u^N(t) = Q^{-1}\tilde{B}'_N e^{(\tau - t)\tilde{A}'_N} \nu, \tag{4.10}$$

$$\nu = \left(\int_0^\tau e^{s\tilde{A}_N} \tilde{B}_N Q^{-1} \tilde{B}'_N e^{s\tilde{A}'_N} \, ds \right)^{-1} (\tilde{x}_N^1 - e^{\tau \tilde{A}_N} \tilde{x}_N^0),$$

where the prime stands for the transpose. Theorem 4.1 implies that the proof of the approximate controllability can be reduced to checking conditions (4.3) and (4.5) with a family of smooth controls $u^N(t)$ given by (4.10).

4.2 Controllability of a Family of Oscillators Without Damping

To show that the conditions given in Theorem 4.1 are not very restrictive, we apply a family of L^2-minimal controls to satisfy estimate (4.5) in the sequel. We will justify this approach for a class distributed parameter systems governed by the following differential equation

$$\dot{x} = Ax + Bu, \quad x \in \ell^2, \ u \in \mathbb{R}. \tag{4.11}$$

We represent elements of the real Hilbert space ℓ^2 as infinite vectors

$$x = (\xi_0, \eta_0, \ \xi_1, \eta_1, \ \xi_2, \eta_2, \ldots)^T$$

and define the norm as

$$\|x\| = \left(\sum_{n=0}^\infty (\xi_n^2 + \eta_n^2) \right)^{1/2}.$$

The linear operator $A : D(A) \to \ell^2$ in (4.11) is given by the following block diagonal matrix

$$A = \begin{pmatrix} A_0 & 0 & 0 & \cdots \\ 0 & A_1 & 0 & \cdots \\ 0 & 0 & A_2 & \cdots \\ \vdots & \vdots & \vdots & \ddots \end{pmatrix},$$

$$A_0 = \begin{pmatrix} 0 & 1 \\ 0 & 0 \end{pmatrix}, \quad A_n = \begin{pmatrix} 0 & \omega_n \\ -\omega_n & 0 \end{pmatrix}, \quad n = 1, 2, \ldots,$$

and

$$
B = \begin{pmatrix} 0 \\ 1 \\ 0 \\ b_1 \\ 0 \\ b_2 \\ \vdots \end{pmatrix} \in \ell^2,
$$

where ω_n and b_n are real constants.

Control system (4.11) is the linear approximation of system (3.23) for the case of a single beam ($k = 1$) and

$$
\omega_n = c_1 \sqrt{\lambda_{1n}}, \quad b_n = -J_{1n}/\rho_1, \quad n = 1, 2, \ldots. \tag{4.12}
$$

To study the controllability problem, we fix a number $N \geq 1$ and consider the following finite dimensional approximation of control system (4.11):

$$
\begin{aligned}
\dot{\xi}_0 &= \eta_0, \\
\dot{\eta}_0 &= u, \\
\dot{\xi}_j &= \omega_j \eta_j, \\
\dot{\eta}_j &= -\omega_j \xi_j + b_j u, \quad j = \overline{1, N}.
\end{aligned} \tag{4.13}
$$

4.2.1 Canonical Form of a Finite-Dimensional Subsystem

In this subsection, we investigate the problem of transforming system (4.13) to the Brunovsky canonical form [14]. Such a transformation is used to parameterize the trajectories by means of an auxiliary function—flat output and its derivatives [15, 16].

A flat output is constructed in the paper [17] for a finite family of oscillators with non-zero frequencies. In contrast to such result, we will consider here the case of system (4.13) whose matrix has a pair of zero eigenvalues and provide a parametrization of the state vector and control explicitly. This parametrization will be used to solve the steering problem and to reduce optimality conditions to the Lagrange problem in the calculus of variations.

Before proving the main result, we formulate an auxiliary lemma.

Lemma 4.1 *Assume that all ω_j^2 are distinct and non-zero for $j = \overline{1, N}$. Then the matrix*

$$W = \begin{pmatrix} 1 & 1 & \cdots & 1 \\ -\omega_1^2 & -\omega_2^2 & \cdots & -\omega_N^2 \\ \omega_1^4 & \omega_2^4 & \cdots & \omega_N^4 \\ \vdots & \vdots & \ddots & \vdots \\ (-\omega_1^2)^{N-1} & (-\omega_2^2)^{N-1} & \cdots & (-\omega_N^2)^{N-1} \end{pmatrix} \qquad (4.14)$$

is non-degenerate, and the elements of W^{-1} are given by the following formula

$$(W^{-1})_{kp} = \left(\prod_{i \neq k} \frac{1}{\omega_i^2 - \omega_k^2} \right) \sum_{(j_1, \ldots, j_{N-p}) \in S_{kp}} (\omega_{j_1} \omega_{j_2} \cdots \omega_{j_{N-p}})^2, \qquad (4.15)$$

$$S_{kp} = \{(j_1, j_2, \ldots, j_{N-p}) : 1 \leq j_1 < j_2 < \cdots < j_{N-p} \leq N, \ j_1, j_2, \ldots, j_{N-p} \neq k\}.$$

If $p = N$ then the sum in (4.15) is assumed to be equal to 1.

The assertion of this Lemma follows from the well-known properties of the Vandermonde matrix (see, e.g., [18]).

Transformation to the Brunovsky canonical form. We introduce the following notations:

$$\xi = \begin{pmatrix} \xi_1 \\ \xi_2 \\ \vdots \\ \xi_N \end{pmatrix} \in \mathbb{R}^N, \quad \eta = \begin{pmatrix} \eta_1 \\ \eta_2 \\ \vdots \\ \eta_N \end{pmatrix} \in \mathbb{R}^N, \quad x = \begin{pmatrix} \xi_0 \\ \eta_0 \\ \xi_1 \\ \eta_1 \\ \vdots \\ \xi_N \\ \eta_N \end{pmatrix} \in \mathbb{R}^{2N+2}.$$

Let us denote by $\tilde{\xi}, \tilde{\eta}$, and $\tilde{x}$ the columns with elements $\tilde{\xi}_j$ and $\tilde{\eta}_j$ in the place of ξ_j and η_j, respectively.

Theorem 4.2 Let $b_j \neq 0$ and let all $\omega_j^2 \neq 0$ be distinct for $j = \overline{1, N}$. Then there exists a non-degenerate linear transformation

$$\tilde{x} = \Phi x, \quad \tilde{v} = \alpha v + \beta \xi, \quad (|\Phi| \neq 0, \ \alpha \neq 0), \qquad (4.16)$$

that brings system (4.13) to the Brunovsky canonical form:

$$\begin{aligned} \dot{\tilde{\xi}}_{j-1} &= \tilde{\eta}_{j-1}, \\ \dot{\tilde{\eta}}_{j-1} &= \tilde{\xi}_j, \quad (j = \overline{1, N}), \\ \dot{\tilde{\xi}}_N &= \tilde{\eta}_N, \\ \dot{\tilde{\eta}}_N &= \tilde{v}. \end{aligned} \qquad (4.17)$$

The components of transformation (4.16) are given by formulas

$$\Phi : \ \tilde{\xi}_0 = c_0\xi_0 + \bar{c}\xi, \ \tilde{\eta}_0 = c_0\eta_0 + \bar{c}\Omega\eta, \ \tilde{\xi} = -WC\Omega^2\xi, \ \tilde{\eta} = -WC\Omega^3\eta,$$

$$(4.18)$$

$$\alpha = c_0 \sum_{j=1}^{N} \left(\omega_j^{2N} \prod_{i \neq j} \frac{\omega_i^2}{\omega_j^2 - \omega_i^2} \right), \quad \beta = (-1)^{N+1}\bar{c}\Omega^{2N+2}, \qquad (4.19)$$

where

$$\bar{c} = (c_1, c_2, \ldots, c_N), \quad C = \begin{pmatrix} c_1 & 0 & \ldots & 0 \\ 0 & c_2 & \ldots & 0 \\ \vdots & \vdots & \ddots & \vdots \\ 0 & 0 & \ldots & c_N \end{pmatrix}, \quad \Omega = \begin{pmatrix} \omega_1 & 0 & \ldots & 0 \\ 0 & \omega_2 & \ldots & 0 \\ \vdots & \vdots & \ddots & \vdots \\ 0 & 0 & \ldots & \omega_N \end{pmatrix}, \quad (4.20)$$

$$c_j = -\frac{c_0}{\omega_j b_j} \prod_{i \neq j} \frac{\omega_i^2}{\omega_i^2 - \omega_j^2}, \quad (j = \overline{1, N}). \qquad (4.21)$$

The matrix W is defined in (4.14), and $c_0 \neq 0$ is an arbitrary constant.

Proof Let us rewrite system (4.13) in the matrix form:

$$\dot{x} = Ax + bv,$$

$$A = \begin{pmatrix} 0 & 1 & 0 & 0 & \ldots & 0 & 0 \\ 0 & 0 & 0 & 0 & \ldots & 0 & 0 \\ 0 & 0 & 0 & \omega_1 & \ldots & 0 & 0 \\ 0 & 0 & -\omega_1 & 0 & \ldots & 0 & 0 \\ \vdots & \vdots & \vdots & \vdots & \ddots & \vdots & \vdots \\ 0 & 0 & 0 & 0 & \ldots & 0 & \omega_N \\ 0 & 0 & 0 & 0 & \ldots & -\omega_N & 0 \end{pmatrix}, \quad b = \begin{pmatrix} 0 \\ 1 \\ 0 \\ b_1 \\ \vdots \\ 0 \\ b_N \end{pmatrix}.$$

In order to transform the system to its canonical form [14], we first find a vector

$$h = (c_0, d_0, c_1, d_1, \ldots, c_N, d_N)$$

such that

$$hb = hAb = hA^2b = \cdots = hA^{2N}b = 0, \quad hA^{2N+1}b = \alpha \neq 0,$$

i.e.

$$\Phi b = \begin{pmatrix} 0 \\ 0 \\ \vdots \\ \alpha \end{pmatrix}, \quad \Phi = \begin{pmatrix} h \\ hA \\ \vdots \\ hA^{2N+1} \end{pmatrix}. \tag{4.22}$$

We assume $d_0 = d_1 = \cdots = d_N = 0$ and compute the matrix Φ:

$$\Phi = \begin{pmatrix} c_0 & 0 & c_1 & 0 & \cdots & c_N & 0 \\ 0 & c_0 & 0 & c_1\omega_1 & \cdots & 0 & c_N\omega_N \\ 0 & 0 & -c_1\omega_1^2 & 0 & \cdots & -c_N\omega_N^2 & 0 \\ 0 & 0 & 0 & -c_1\omega_1^3 & \cdots & 0 & -c_N\omega_N^3 \\ \vdots & \vdots & \vdots & \vdots & \ddots & \vdots & \vdots \\ 0 & 0 & (-1)^N c_1\omega_1^{2N} & 0 & \cdots & (-1)^N c_N\omega_N^{2N} & 0 \\ 0 & 0 & 0 & (-1)^N c_1\omega_1^{2N+1} & \cdots & 0 & (-1)^N c_N\omega_N^{2N+1} \end{pmatrix}. \tag{4.23}$$

It is easy to see that $hA^{2j}b = 0$ for all $j = \overline{0, N}$ with the above choice of $d_0, d_1,\ldots,$ d_N. By using the even rows of Φ, we conclude that the conditions $hA^{2j+1}b = 0$, $j = \overline{0, N-1}$ are equivalent to the following system of linear algebraic equations with respect to $c_0, c_1,\ldots, c_N$:

$$W \begin{pmatrix} b_1\omega_1 c_1 \\ b_2\omega_2 c_2 \\ \vdots \\ b_N\omega_N c_N \end{pmatrix} = \begin{pmatrix} -c_0 \\ 0 \\ \vdots \\ 0 \end{pmatrix}.$$

Application of the inverse matrix (4.15) to this system yields formula (4.21) for the coefficients $c_1, c_2,\ldots, c_N$, depending on the value of c_0. Let us show that, for any $c_0 \neq 0$, the condition $hA^{2N+1}b \neq 0$ holds in (4.22):

$$hA^{2N+1}b = (-1)^N \sum_{j=1}^{N} b_j c_j \omega_j^{2N+1} = (-1)^{N-1} c_0 \sum_{j=1}^{N} \omega_j^{2N} \prod_{i\neq j} \frac{\omega_i^2}{\omega_i^2 - \omega_j^2}$$

$$= c_0 \sum_{j=1}^{N} \omega_j^{2N} \prod_{i\neq j} \frac{\omega_i^2}{\omega_j^2 - \omega_i^2}, \tag{4.24}$$

so that the obtained expression coincides with $\alpha \neq 0$ in (4.19).

Let us define $\tilde{x} = \Phi x$. By taking into account the structure of the matrix (4.23) and notations (4.20), we conclude that the components of the map $x \mapsto \tilde{x}$ satisfy formula (4.18). Let us show that Φ is non-degenerate:

$$|\Phi| = c_0^2 \begin{vmatrix} -c_1\omega_1^2 & 0 & \cdots & -c_N\omega_N^2 & 0 \\ 0 & -c_1\omega_1^3 & \cdots & 0 & -c_N\omega_N^3 \\ c_1\omega_1^4 & 0 & \cdots & c_N\omega_N^4 & 0 \\ 0 & c_1\omega_1^5 & \cdots & 0 & c_N\omega_N^5 \\ \vdots & \vdots & \ddots & \vdots & \vdots \\ (-1)^N c_1\omega_1^{2N} & 0 & \cdots & (-1)^N c_N\omega_N^{2N} & 0 \\ 0 & (-1)^N c_1\omega_1^{2N+1} & \cdots & 0 & (-1)^N c_N\omega_N^{2N+1} \end{vmatrix}$$

$$= (c_0 c_1 \cdots c_N)^2 (\omega_1\omega_2\cdots\omega_N)^5 |W|^2$$

$$= (c_0 c_1 \cdots c_N)^2 (\omega_1\omega_2\cdots\omega_N)^5 \prod_{1\le i<j\le N} (\omega_j^2 - \omega_i^2)^2.$$

Thus, $|\Phi| \ne 0$ under the assumptions of the Theorem. By differentiating $\tilde{x}$ along the trajectories of system (4.13), we get:

$$\dot{\tilde{\xi}}_j = \frac{d}{dt}(hA^{2j}x) = hA^{2j+1}x + hA^{2j}bv = hA^{2j+1}x = \tilde{\eta}_j, \quad (j = \overline{0, N}),$$

$$\dot{\tilde{\eta}}_j = \frac{d}{dt}(hA^{2j+1}x) = hA^{2j+2}x + hA^{2j+1}bv = hA^{2j+2}x = \tilde{\xi}_{j+1}, \quad (j = \overline{0, N-1}),$$

$$\dot{\tilde{\eta}}_N = \frac{d}{dt}(hA^{2N+1}x) = hA^{2N+2}x + hA^{2N+1}bv. \tag{4.25}$$

As

$$hA^{2N+2}x = (-1)^{N+1} \sum_{j=1}^{N} c_j \omega_j^{2N+2}\xi_j,$$

then (4.25) together with (4.19) and (4.24) imply $\dot{\tilde{\eta}}_N = \alpha v + \beta\xi = \tilde{v}$. Therefore, system (4.13) takes the form (4.17) in variables $\tilde{x}$ and $\tilde{v}$.

Theorem 4.3 *Let the assumptions of Theorem 4.2 be satisfied. Then for each*

$$t_0 < t_1, \quad x^0 = \begin{pmatrix} \xi_0^0 \\ \eta_0^0 \\ \vdots \\ \xi_N^0 \\ \eta_N^0 \end{pmatrix} \in \mathbb{R}^{2N+2}, \quad x^1 = \begin{pmatrix} \xi_0^1 \\ \eta_0^1 \\ \vdots \\ \xi_N^1 \\ \eta_N^1 \end{pmatrix} \in \mathbb{R}^{2N+2},$$

there exists a control $v = v_{x^0 x^1}(t)$ *for which system (4.13) has the solution* $x(t)$, $t \in [t_0, t_1]$, *that satisfies the boundary conditions* $x(t_0) = x^0$ *and* $x(t_1) = x^1$. *Such control is defined by the formula*

$$v_{x^0 x^1}(t) = \alpha^{-1} \left(\frac{d^{2N+2}}{dt^{2N+2}} + \sum_{p=1}^{N} \gamma_p \frac{d^{2p}}{dt^{2p}} \right) y(t), \quad t \in [t_0, t_1], \qquad (4.26)$$

where $y \in C^{2N+2}[t_0, t_1]$ *is an arbitrary function satisfying the boundary conditions:*

$$y(t_i) = \sum_{j=0}^{N} c_j \xi_j^i, \quad \dot{y}(t_i) = c_0 \eta_0^i + \sum_{j=1}^{N} c_j \omega_j \eta_j^i,$$

$$\begin{pmatrix} \frac{d^2}{dt^2} y \\ \frac{d^4}{dt^4} y \\ \vdots \\ \frac{d^{2N}}{dt^{2N}} y \end{pmatrix}_{t=t_i} = -WC\Omega^2 \begin{pmatrix} \xi_1^i \\ \xi_2^i \\ \vdots \\ \xi_N^i \end{pmatrix},$$

$$\begin{pmatrix} \frac{d^3}{dt^3} y \\ \frac{d^5}{dt^5} y \\ \vdots \\ \frac{d^{2N+1}}{dt^{2N+1}} y \end{pmatrix}_{t=t_i} = -WC\Omega^3 \begin{pmatrix} \eta_1^i \\ \eta_2^i \\ \vdots \\ \eta_N^i \end{pmatrix}, \quad i = 0, 1, \qquad (4.27)$$

$$\gamma_p = \sum_{k=1}^{N} \left\{ \omega_k^{2N} \left(\prod_{\substack{1 \le l \le N \\ l \ne k}} \frac{1}{\omega_k^2 - \omega_l^2} \right) \sum_{\substack{1 \le j_1 < j_2 < \cdots < j_{N-p} \le N \\ j_1, j_2, \ldots, j_{N-p} \ne k}} (\omega_{j_1} \omega_{j_2} \cdots \omega_{j_{N-p}})^2 \right\}. $$

$$(4.28)$$

In formula (4.28), the sum for $(j_1, \ldots, j_{N-p})$ *is assumed to be 1 if* $p = N$; *all products with the empty set of indices are assumed to be 1 as well.*

Proof Let us transform system (4.13) to the Brunovsky canonical form (4.17) by using the change of variables (4.16). If $y(t)$ is a function of class $C^{2N+2}[t_0, t_1]$ and the boundary conditions (4.27) hold, then

$$\tilde{x}(t) = \left(y(t), \frac{d}{dt} y(t), \ldots, \frac{d^{2N+1}}{dt^{2N+1}} y(t) \right)^T$$

is a solution of system (4.17) with the control

$$\tilde{v}(t) = \frac{d^{2N+2}}{dt^{2N+2}} y(t).$$

According to formulas (4.16), the respective function $x(t) = \Phi^{-1}\tilde{x}(t)$ satisfies the boundary conditions

$$x(t_0) = x^0, \quad x(t_1) = x^1$$

and system (4.13) with the control

$$v(t) = \alpha^{-1}(\tilde{v}(t) - \beta\xi(t)). \tag{4.29}$$

By using the property $\ddot{\tilde{\xi}} = -WC\Omega^2\xi$ in (4.18), we rewrite the relation (4.29) as follows:

$$v(t) = \alpha^{-1}\frac{d^{2N+2}}{dt^{2N+2}}y(t) + \alpha^{-1}\beta\Omega^{-2}C^{-1}W^{-1}\left(\frac{d^2}{dt^2}y(t), \frac{d^4}{dt^4}y(t), \ldots, \frac{d^{2N}}{dt^{2N}}y(t)\right)^T.$$

Then the application of Lemma 4.1 yields formula (4.26).

The above theorem reduces the control design problem to the construction of a function $y(t)$ that satisfies the set of $2 \times (2N + 2)$ boundary conditions (4.27). In particular, such a function $y(t)$ can be taken as a polynomial. Note that the formula for the control (4.26) exploits only derivatives of even order of $y(t)$.

The result obtained allows us to solve the steering problem for the control system (4.13) explicitly for an arbitrary number of elastic modes. It is of great interest to check the applicability of the flatness based approach for a nonlinear modification of system (4.13). We will treat this issue in the next section.

4.2.2 Feedback Linearization Test

A system of differential equations

$$\dot{x} = F(x, u), \quad x \in \mathbb{R}^n, \quad u \in \mathbb{R}^m \tag{4.30}$$

is said to be *flat* [15] if there exist numbers $k, l \geq 0$ and functions

$$h : \mathbb{R}^n \times \left(\mathbb{R}^m\right)^{k+1} \to \mathbb{R}^m, \quad \varphi : \left(\mathbb{R}^m\right)^{l+1} \to \mathbb{R}^n, \quad \psi : \left(\mathbb{R}^m\right)^{l+1} \to \mathbb{R}^m,$$

such that

$$y = h(x, u, \dot{u}, \ldots, u^{(k)}), \quad x = \varphi(y, \dot{y}, \ldots, y^{(l)}), \quad u = \psi(y, \dot{y}, \ldots, y^{(l)}). \tag{4.31}$$

Here x is the state vector and u is the control of system (4.30). It is assumed that functions F, h, φ, and ψ satisfy certain regularity properties. It is a well-known fact that the question of the existence of such a function h (called "flat output") remains open for general nonlinear systems [19]. However, it is possible to use efficient algebraic techniques for checking the flatness of some specific classes of systems.

Consider a particular case when the system (4.30) is control affine and $m = 1$:

$$\dot{x} = f(x) + g(x)u, \quad x = (x_1, x_2, \ldots, x_n)^T \in \mathbb{R}^n, \quad u \in \mathbb{R}^1, \tag{4.32}$$

where $f, g \in C^\infty(\mathbb{R}^n)$. System (4.32) is said to be *linearizable* (by a static feedback law), if there is a diffeomorphism

$$\tilde{x} = \Phi(x) \tag{4.33}$$

and a feedback transformation

$$\tilde{u} = \alpha(x)u + \beta(x), \quad \beta(x) \neq 0, \tag{4.34}$$

such that system (4.32) is transformed to the Brunovsky canonical form:

$$\dot{\tilde{x}}_1 = \tilde{x}_2, \quad \dot{\tilde{x}}_2 = \tilde{x}_3, \ldots, \dot{\tilde{x}}_{n-1} = \tilde{x}_n, \quad \dot{\tilde{x}}_n = \tilde{u}.$$

Here $\tilde{x} = (\tilde{x}_1, \tilde{x}_2, \ldots, \tilde{x}_n)^T$ is the new state and $\tilde{u}$ is the new control. If the above transformation (4.33) and (4.34) exists for the values of x in some neighborhood of a point $x_0 \in \mathbb{R}^n$, then system (4.32) is called *locally linearizable in a neighborhood of x_0*. According to the paper [20], the existence of a flat output h and functions φ, ψ for system (4.32) is equivalent to linearizability by means of a static feedback law. In order to formulate the linearizability conditions, we introduce the Lie bracket of vector fields f and g:

$$[f, g](x) = \frac{\partial g(x)}{\partial x} f(x) - \frac{\partial f(x)}{\partial x} g(x),$$

here $\frac{\partial f(x)}{\partial x}$ and $\frac{\partial g(x)}{\partial x}$ are the Jacobian matrices of f and g, respectively. Let us also consider the following iterated Lie brackets:

$$ad_f^0 g = g, \quad ad_f^1 g = [f, g], \quad ad_f^{j+1} g = [f, ad_f^j g], \quad j = 1, 2, \ldots$$

Proposition 4.1 [21, 22] *System (4.32) is locally linearizable in a neighborhood of $x_0 \in \mathbb{R}^n$ iff the following conditions are satisfied in some neighborhood G of x_0:*

(C1) *the vector fields $\{g(x), ad_f g(x), \ldots, ad_f^{n-1} g(x)\}$ are linearly independent for each $x \in G$;*

(C2) *there exist smooth functions $\gamma_q^{ij}(x), 0 \leq i, j \leq n - 2$, such that*

$$[ad_f^i g, ad_f^j g](x) = \sum_{q=0}^{n-2} \gamma_q^{ij}(x) ad_f^q g(x), \quad \forall x \in G.$$

Rotation of a body with a flexible beam. Let us consider a finite dimensional approximation of system (3.22) with modes of numbers up to N for the case of a single beam ($k = 1$):

$$
\begin{aligned}
\dot{\xi}_0 &= \eta_0, \\
\dot{\eta}_0 &= v, \\
\dot{\xi}_j &= \omega_j \eta_j, \\
\dot{\eta}_j &= (\eta_0^2 r_j - \omega_j)\xi_j + b_j v, \quad (j = \overline{1, N}),
\end{aligned}
\tag{4.35}
$$

here $\xi_0 = \phi$ is the angle of rotation of the rigid body, $\eta_0 = \omega$ its angular velocity, $v \in \mathbb{R}$ is the control. Let us recall that ξ_j and η_j correspond to the jth modal coordinate and velocity, respectively.

We will prove that system (4.35) is not linearizable (by a static feedback law) for any number $N \geq 1$ if $r_1, r_2, \ldots, r_N \neq 0$.

Theorem 4.4 *Let $0 < \omega_1 < \cdots < \omega_N$, $b_j \neq 0$, $r_j \neq 0$ for all $j = \overline{1, N}$. Then system (4.35) is not linearizable by a static feedback law.*

Proof Let us denote

$$
z = \begin{pmatrix} \xi_0 \\ \eta_0 \\ \xi \\ \eta \end{pmatrix}, \quad
\xi = \begin{pmatrix} \xi_1 \\ \xi_2 \\ \vdots \\ \xi_N \end{pmatrix}, \quad
\eta = \begin{pmatrix} \eta_1 \\ \eta_2 \\ \vdots \\ \eta_N \end{pmatrix}
$$

and rewrite system (4.35) in the following form:

$$
\dot{z} = f(z) + g(z)v, \quad z \in \mathbb{R}^{2N+2}, \quad v \in \mathbb{R}, \tag{4.36}
$$

where z is the state and v is the control,

$$
f(z) = \begin{pmatrix} F_0 & 0_{2 \times N} & 0_{2 \times N} \\ 0_{N \times 2} & 0_{N \times N} & \Omega \\ 0_{N \times 2} & \eta_0^2 R - \Omega & 0_{N \times N} \end{pmatrix} z, \quad
g(z) = \begin{pmatrix} 0 \\ 1 \\ 0_{N \times 1} \\ b \end{pmatrix}, \quad
F_0 = \begin{pmatrix} 0 & 1 \\ 0 & 0 \end{pmatrix},
$$

$$
\Omega = \begin{pmatrix} \omega_1 & 0 & \cdots & 0 \\ 0 & \omega_2 & \cdots & 0 \\ \vdots & \vdots & \ddots & \vdots \\ 0 & 0 & \cdots & \omega_N \end{pmatrix}, \quad
R = \begin{pmatrix} r_1 & 0 & \cdots & 0 \\ 0 & r_2 & \cdots & 0 \\ \vdots & \vdots & \ddots & \vdots \\ 0 & 0 & \cdots & r_N \end{pmatrix}, \quad
b = \begin{pmatrix} b_1 \\ b_2 \\ \vdots \\ b_N \end{pmatrix}.
$$

In the above block matrices, $0_{i \times j}$ denotes the block of zeroes with i rows and j columns. We show that condition ($C2$) from Proposition 4.1 is not satisfied for system (4.36) in any neighborhood of $z = 0$. For this purpose, we compute the

Jacobian matrices and Lie brackets as follows:

$$\frac{\partial f(z)}{\partial z} = \begin{pmatrix} 0 & 1 & 0_{1\times N} & 0_{1\times N} \\ 0 & 0 & 0_{1\times N} & 0_{1\times N} \\ 0_{N\times 1} & 0_{N\times 1} & 0_{N\times N} & \Omega \\ 0_{N\times 1} & 2\eta_0 R\xi & \eta_0^2 R - \Omega & 0_{N\times N} \end{pmatrix}, \quad \frac{\partial g(z)}{\partial z} = 0_{(2N+2)\times(2N+2)},$$

$$ad_f^0 g = \begin{pmatrix} 0 \\ 1 \\ 0_{N\times 1} \\ b \end{pmatrix}, \quad ad_f^1 g = [f, g] = -\frac{\partial f(z)}{\partial z} g = -\begin{pmatrix} 1 \\ 0 \\ \Omega b \\ 2\eta_0 R\xi \end{pmatrix}. \qquad (4.37)$$

As the diagonal matrices Ω and R commute, we obtain

$$ad_f^{2j} g = \begin{pmatrix} 0 \\ 0 \\ Q_j(\eta_0)\xi \\ -Q_j(\eta_0)\eta + P_j(\eta_0)b \end{pmatrix}, \quad ad_f^{2j+1} g = \begin{pmatrix} 0 \\ 0 \\ 2Q_j(\eta_0)\Omega\eta - P_j(\eta_0)\Omega b \\ 2Q_j(\eta_0)(\Omega - \eta_0^2 R)\xi \end{pmatrix}$$

$$\qquad (4.38)$$

for all $j \in \mathbb{N}$, where

$$Q_j(\eta_0) = 2^{2j-1}\eta_0 R\Omega^j (\eta_0^2 R - \Omega)^{j-1}, \quad P_j(\eta_0) = \Omega^j (\eta_0^2 R - \Omega)^j. \qquad (4.39)$$

Let us check condition (C2) of Proposition 4.1 for $[ad_f^0 g, ad_f^1 g]$. If condition (C2) is satisfied then, for each z in a neighborhood of the origin, the following relation holds

$$[ad_f^0 g, ad_f^1 g] = \begin{pmatrix} 0_{2\times 1} \\ 0_{N\times 1} \\ -2R\xi \end{pmatrix} = \sum_{j=0}^{2N} \gamma_j ad_f^j g(z) \qquad (4.40)$$

with some $\gamma_j = \gamma_j(z)$. By using expressions (4.37), (4.38) and computing coordinates of the vectors in (4.40), we get $\gamma_0 = \gamma_1 = 0$,

$$\begin{pmatrix} 0_{N\times 1} \\ -2R\xi \end{pmatrix} = \sum_{j=1}^{N} \gamma_{2j} \begin{pmatrix} Q_j\xi \\ -Q_j\eta + P_j b \end{pmatrix} + \sum_{j=1}^{N-1} \gamma_{2j+1} \begin{pmatrix} 2\Omega(Q_j\eta - P_j)b \\ 2Q_j(\Omega - \eta_0^2 R)\xi \end{pmatrix}.$$

$$\qquad (4.41)$$

For a fixed z, relations (4.41) may be considered as a system of $2N$ linear algebraic equations with respect to $2N-1$ variables $\gamma_2, \gamma_3, \ldots, \gamma_{2N}$. According to the Rouché–Capelli theorem, that system is consistent iff the rank of the $2N \times 2N$-matrix

$$M = \begin{pmatrix} 0_{N\times 1} & Q_1\xi & 2\Omega(Q_1\eta - P_1)b & \cdots & 2\Omega(Q_{N-1}\eta - P_{N-1})b & Q_N\xi \\ -2R\xi & -Q_1\eta + P_1 b & 2Q_1(\Omega - \eta_0^2 R)\xi & & 2Q_{N-1}(\Omega - \eta_0^2 R)\xi & -Q_N\eta + P_N b \end{pmatrix}$$

is equal to the rank of its $2N \times (2N - 1)$-submatrix obtained by striking out the first column of M. This implies that if each neighborhood of $z_0 = 0$ contains points z such that rank $M = 2N$ then system (4.41) is not consistent, i.e. condition (C2) is violated. Let $N = 1$ then

$$\det M = \begin{vmatrix} O & Q_1\xi \\ -2R\xi & -Q_1\eta + P_1 b \end{vmatrix} = 2RQ_1\xi_1^2 = 4r_1^2\omega_1\eta_0\xi_1^2,$$

hence, condition (C2) does not hold at each point $z = (\xi_0, \eta_0, \xi_1, \eta_1)^T \in \mathbb{R}^4$ outside of the hyperplanes $\xi_1 = 0$, $\eta_0 = 0$.

Consider now the case $N > 1$. Assume that $\xi_1 = \cdots = \xi_{N-1} = 0$ and compute $\det M$:

$$\det M = (-1)^N 2r_N\xi_N\omega_1\omega_2\cdots\omega_{N-1}$$

$$\times \begin{vmatrix} 0 & \cdots & 0 & 2q_1^1\eta_1 - p_1^1 b_1 & \cdots & 2q_1^{N-1}\eta_1 - p_1^{N-1}b_1 \\ \vdots & \ddots & \vdots & \vdots & \ddots & \vdots \\ 0 & \cdots & 0 & 2q_{N-1}^1\eta_{N-1} - p_{N-1}^1 b_{N-1} & \cdots & 2q_{N-1}^{N-1}\eta_{N-1} - p_{N-1}^{N-1}b_{N-1} \\ q_N^1\xi_N & \cdots & q_N^N\xi_N & \omega_N(2q_N^1\eta_N - p_N^1 b_N) & \cdots & \omega_N(2q_N^{N-1}\eta_N - p_N^{N-1}b_N) \\ p_1^1 b_1 - q_1^1\eta_1 & \cdots & p_1^N b_1 - q_1^N\eta_1 & 0 & \cdots & 0 \\ \vdots & \ddots & \vdots & \vdots & \ddots & \vdots \\ p_{N-1}^1 b_{N-1} - q_{N-1}^1\eta_{N-1} & \cdots & p_{N-1}^N b_{N-1} - q_{N-1}^N\eta_{N-1} & 0 & \cdots & 0 \end{vmatrix},$$

where $(q_1^j, q_2^j, \ldots, q_N^j)$ and $(p_1^j, p_2^j, \ldots, p_N^j)$ are diagonal elements of the matrices $Q_j(\eta_0)$ and $P_j(\eta_0)$, respectively. By expanding this determinant with respect to the minors of the $(N - 1)$th order from the upper rows of M, we get

$$\det M = \pm 2r_N\xi_N^2\omega_1\omega_2\ldots\omega_{N-1}\Delta_1\Delta_2, \qquad (4.42)$$

$$\Delta_1 = \begin{vmatrix} 2q_1^1\eta_1 - p_1^1 b_1 & \cdots & 2q_1^{N-1}\eta_1 - p_1^{N-1}b_1 \\ \vdots & \ddots & \vdots \\ 2q_{N-1}^1\eta_{N-1} - p_{N-1}^1 b_{N-1} & \cdots & 2q_{N-1}^{N-1}\eta_{N-1} - p_{N-1}^{N-1}b_{N-1} \end{vmatrix},$$

$$\Delta_2 = \begin{vmatrix} q_N^1 & \cdots & q_N^N \\ p_1^1 b_1 - q_1^1\eta_1 & \cdots & p_1^N b_1 - q_1^N\eta_1 \\ \vdots & \ddots & \vdots \\ p_{N-1}^1 b_{N-1} - q_{N-1}^1\eta_{N-1} & \cdots & p_{N-1}^N b_{N-1} - q_{N-1}^N\eta_{N-1} \end{vmatrix}.$$

As formulas (4.39) imply

$$2Q_j(\eta_0)\eta - P_j(\eta_0)b = (-1)^{j+1}\Omega^j b + O(\eta_0)$$

for $\eta_0 \to 0$, then we may write Δ_1 as follows:

$$\Delta_1 = \begin{vmatrix} \omega_1 b_1 & -\omega_1^2 b_1 & \cdots & (-1)^N \omega_1^{N-1} b_1 \\ \omega_2 b_2 & -\omega_2^2 b_2 & \cdots & (-1)^N \omega_2^{N-1} b_2 \\ \vdots & \vdots & \ddots & \vdots \\ \omega_{N-1} b_{N-1} & -\omega_{N-1}^2 b_{N-1} & \cdots & (-1)^N \omega_{N-1}^{N-1} b_{N-1} \end{vmatrix} + O(\eta_0)$$

$$= (\omega_1 b_1)(\omega_2 b_2) \ldots (\omega_{N-1} b_{N-1}) \prod_{1 \le i < j \le N-1} (\omega_i - \omega_j) + O(\eta_0). \qquad (4.43)$$

In expression (4.43), we use a well-known formula for the Vandermonde determinant. By exploiting (4.39), we get

$$Q_j(\eta_0) = 2\eta_0 \Omega R(-4\Omega^2)^{j-1} + O(\eta_0^2), \quad P_j(\eta_0)b - Q_j(\eta_0)\eta = (-\Omega^2)^j b + O(\eta_0).$$

Hence,

$$\Delta_2 = 2\eta_0 r_N \omega_N \begin{vmatrix} 1 & -4\omega_N^2 & \cdots & (-4\omega_N^2)^{N-1} \\ -\omega_1^2 b_1 & (-\omega_1^2)^2 b_1 & \cdots & (-\omega_1^2)^N b_1 \\ \vdots & \vdots & \ddots & \vdots \\ -\omega_{N-1}^2 b_{N-1} & (-\omega_{N-1}^2)^2 b_{N-1} & \cdots & (-\omega_{N-1}^2)^N b_{N-1} \end{vmatrix} + O(\eta_0^2)$$

$$= 2\eta_0 r_N \omega_N \left(\prod_{i=1}^{N-1} b_i \omega_i^2 (\omega_i^2 - 4\omega_N^2) \right) \prod_{1 \le i < j \le N-1} (\omega_i^2 - \omega_j^2) + O(\eta_0^2). \qquad (4.44)$$

Expressions (4.42), (4.43), and (4.44) imply that rank $M = 2N$ if $\eta_0 \ne 0$ and $\xi_N \ne 0$ are small enough. It means that condition (C2) of Proposition 4.1 does not hold in any neighborhood of $z_0 = 0$. Therefore, system (4.36) is not locally linearizable in a neighborhood of $z_0 = 0$. As $m = \dim v = 1$ the system (4.36) is not flat because of the result of [20].

The above result confirms that Theorem 4.2 on the transformation of the linear system (i.e. system (4.35) with $r_1 = \cdots = r_N = 0$) to the Brunovsky canonical form is final, in the sense that its nonlinear perturbation (i.e. system (4.35) with $r_1, \ldots, r_N \ne 0$) is not flat. This conclusion is valid for an arbitrary finite number N of modal coordinates.

4.2.3 L^2-Minimal Control

In this subsection, we present a result on the optimal control for finite-dimensional system (4.13).

Theorem 4.5 *Let $\tau > 0$,*

$$
x_N^0 = \begin{pmatrix} \xi_0^0 \\ \eta_0^0 \\ \vdots \\ \xi_N^0 \\ \eta_N^0 \end{pmatrix} \in \mathbb{R}^{2N+2}, \quad x_N^1 = \begin{pmatrix} \xi_0^1 \\ \eta_0^1 \\ \vdots \\ \xi_N^1 \\ \eta_N^1 \end{pmatrix} \in \mathbb{R}^{2N+2}.
$$

Then there exists a unique $u = u_{x^0,x^1}^N(t)$ that solves the optimal control problem for system (4.13) with boundary conditions

$$
\xi_j(0) = \xi_j^0, \quad \eta_j(0) = \eta_j^0, \quad \xi_j(\tau) = \xi_j^1, \quad \eta_j(\tau) = \eta_j^1, \quad j = \overline{0, N}, \tag{4.45}
$$

and the cost

$$
J(u) = \int_0^\tau |u(t)|^2 \, dt \rightarrow \min \tag{4.46}
$$

in the class of admissible controls $u \in L^2(0, \tau)$. This optimal control is given by the formula

$$
u_{x^0,x^1}^N(t) = k_0 + k_1 t + \sum_{j=1}^N \left(U_j \cos(\omega_j t) + V_j \sin(\omega_j t) \right), \tag{4.47}
$$

where the coefficients k_0, k_1, U_j, V_j satisfy the following system of linear algebraic equations:

$$
(M + F) \begin{pmatrix} k_0 \\ \tau k_1 \\ U_1 \\ V_1 \\ \vdots \\ U_N \\ V_N \end{pmatrix} = \begin{pmatrix} (\xi_0^1 - \xi_0^0)/\tau - \eta_0^0 \\ \eta_0^1 - \eta_0^0 \\ (\xi_1^1 \sin \omega_1 \tau + \eta_1^1 \cos \omega_1 \tau - \eta_1^0)/b_1 \\ (\xi_1^1 \cos \omega_1 \tau - \eta_1^1 \sin \omega_1 \tau - \xi_1^0)/b_1 \\ \vdots \\ (\xi_N^1 \sin \omega_N \tau + \eta_N^1 \cos \omega_N \tau - \eta_N^0)/b_N \\ (\xi_N^1 \cos \omega_N \tau - \eta_N^1 \sin \omega_N \tau - \xi_N^0)/b_N \end{pmatrix}, \tag{4.48}
$$

$$
M = \begin{pmatrix} \begin{pmatrix} \frac{\tau}{2} & \frac{\tau}{2} \\ \tau & \frac{\tau}{2} \end{pmatrix} & O & \cdots & O \\ O & \begin{pmatrix} \frac{\tau}{0} & 0 \\ 0 & -\frac{\tau}{2} \end{pmatrix} & \cdots & O \\ \vdots & \vdots & \ddots & \vdots \\ O & O & \cdots & \begin{pmatrix} \frac{\tau}{0} & 0 \\ 0 & -\frac{\tau}{2} \end{pmatrix} \end{pmatrix}. \tag{4.49}
$$

Here the components of matrix $F = (f_{lk})$ are as follows

$$f_{1,2i+1} = \frac{1 - \cos(\omega_i\tau)}{\omega_i^2\tau}, \quad f_{1,2i+2} = \frac{\omega_i\tau - \sin(\omega_i\tau)}{\omega_i^2\tau},$$

$$f_{2,2i+1} = \frac{\sin(\omega_i\tau)}{\omega_i}, \quad f_{2,2i+2} = \frac{1 - \cos(\omega_i\tau)}{\omega_i},$$

$$f_{2j+1,1} = \frac{\sin(\omega_j\tau)}{\omega_j}, \quad f_{2j+1,2} = \frac{\omega_j\tau\sin(\omega_j\tau) + \cos(\omega_j\tau) - 1}{\omega_j^2\tau},$$

$$f_{2j+1,2j+1} = \frac{\sin(2\omega_j\tau)}{4\omega_j}, \quad f_{2j+1,2j+2} = \frac{\sin^2(\omega_j\tau)}{2\omega_j},$$

$$f_{2j+1,2i+1} = \frac{1}{2}\left(\frac{\sin(\omega_i + \omega_j)\tau}{\omega_i + \omega_j} + \frac{\sin(\omega_i - \omega_j)\tau}{\omega_i - \omega_j}\right),$$

$$f_{2j+1,2i+2} = \frac{1}{2}\left(\frac{2\omega_i}{\omega_i^2 - \omega_j^2} + \frac{\cos(\omega_i - \omega_j)\tau}{\omega_j - \omega_i} - \frac{\cos(\omega_i + \omega_j)\tau}{\omega_i + \omega_j}\right),$$

$$f_{2j+2,1} = \frac{\cos(\omega_j\tau) - 1}{\omega_j},$$

$$f_{2j+2,2} = \frac{\omega_j\tau\cos(\omega_j\tau) - \sin(\omega_j\tau)}{\omega_j^2\tau}, \quad f_{2j+2,2j+1} = -\frac{\sin^2(\omega_j\tau)}{2\omega_j},$$

$$f_{2j+2,2j+2} = \frac{\sin(2\omega_j\tau)}{4\omega_j},$$

$$f_{2j+2,2i+1} = \frac{1}{2}\left(\frac{\cos(\omega_i + \omega_j)\tau}{\omega_i + \omega_j} + \frac{\cos(\omega_j - \omega_i)\tau}{\omega_j - \omega_i} + \frac{2\omega_j}{\omega_i^2 - \omega_j^2}\right),$$

$$f_{2j+2,2i+2} = \frac{1}{2}\left(\frac{\sin(\omega_i + \omega_j)\tau}{\omega_i + \omega_j} + \frac{\sin(\omega_i - \omega_j)\tau}{\omega_j - \omega_i}\right), \quad i, j = \overline{1, N}, \quad i \neq j.$$

$$(4.50)$$

For proving this result we need some auxiliary constructions.

Under the conditions of Theorem 4.5, each continuous control $v(t)$ that solves the steering problem for system (4.13) with boundary conditions (4.45) and (4.46) may be presented in the form (4.26) by Theorem 4.3, where α and γ_p are defined by formulas (4.19) and (4.28), and the function $y \in C^{2N+2}[t_0, t_1]$ satisfies the boundary conditions:

$$y(t_i) = \sum_{j=0}^{N} c_j\xi_j^i, \quad \dot{y}(t_i) = c_0\eta_0^i + \sum_{j=1}^{N} c_j\omega_j\eta_j^i,$$

$$
\begin{pmatrix}
\frac{d^2}{dt^2} y \\
\frac{d^4}{dt^4} y \\
\vdots \\
\frac{d^{2N}}{dt^{2N}} y
\end{pmatrix}_{t=t_i}
= -W
\begin{pmatrix}
c_1 \omega_1^2 \varsigma_1^i \\
c_2 \omega_2^2 \varsigma_2^i \\
\vdots \\
c_N \omega_N^2 \varsigma_N^i
\end{pmatrix},
$$

$$
\begin{pmatrix}
\frac{d^3}{dt^3} y \\
\frac{d^5}{dt^5} y \\
\vdots \\
\frac{d^{2N+1}}{dt^{2N+1}} y
\end{pmatrix}_{t=t_i}
= -W
\begin{pmatrix}
c_1 \omega_1^3 \eta_1^i \\
c_2 \omega_2^3 \eta_2^i \\
\vdots \\
c_N \omega_N^3 \eta_N^i
\end{pmatrix},
\quad i = 0, 1. \tag{4.51}
$$

The matrix W is defined in (4.14),

$$
c_j = -\frac{c_0}{\omega_j b_j} \prod_{\substack{1 \le i \le N \\ i \ne j}} \frac{\omega_i^2}{\omega_i^2 - \omega_j^2}.
$$

Here $c_0 \ne 0$ is an arbitrary constant. We assume that all products with respect to the empty set of indices are equal to one. We present a more convenient form of control (4.26) in the following lemma.

Lemma 4.2 *Formula (4.26) is equivalent to the following one:*

$$
v(t) = \frac{1}{c_0 \omega_1^2 \omega_2^2 \dots \omega_N^2} \left(\frac{d^2}{dt^2} + \omega_1^2 \right) \left(\frac{d^2}{dt^2} + \omega_2^2 \right) \cdots \left(\frac{d^2}{dt^2} + \omega_N^2 \right) \frac{d^2 y(t)}{dt^2}, \quad t \in [t_0, t_1]. \tag{4.52}
$$

Proof By expanding formula (4.52) and comparing the coefficients at derivatives of $y(t)$ in (4.26) and (4.52), we get:

$$
\alpha = c_0 \omega_1^2 \omega_2^2 \dots \omega_N^2, \quad \gamma_p = \sum_{1 \le j_1 < j_2 < \cdots < j_{N-p+1} \le N} \omega_{j_1}^2 \omega_{j_2}^2 \dots \omega_{j_{N-p+1}}^2. \tag{4.53}
$$

To prove this lemma, it suffices to establish properties (4.53), where α and γ_p are defined by expressions (4.19) and (4.28). For this purpose we compute the determinant of the matrix with rows of W from formula (4.14):

$$
\begin{vmatrix}
1 & 1 & \cdots & 1 \\
-\omega_1^2 & -\omega_2^2 & \cdots & -\omega_N^2 \\
(-\omega_1^2)^2 & (-\omega_2^2)^2 & \cdots & (-\omega_N^2)^2 \\
\vdots & \vdots & \ddots & \vdots \\
(-\omega_1^2)^{N-2} & (-\omega_2^2)^{N-2} & \cdots & (-\omega_N^2)^{N-2} \\
(-\omega_1^2)^{N-m} & (-\omega_2^2)^{N-m} & \cdots & (-\omega_N^2)^{N-m}
\end{vmatrix}
= \sum_{k=1}^{N} (-1)^{k+m} \omega_k^{2(N-m)} \Delta_k, \tag{4.54}
$$

where $1 \le m \le N$ and Δ_k is the minor obtained by cancelling the last row and k-th column of W. Formula (4.54) is obtained by the expansion of the determinant with respect to the last row. A well-known formula for the Vandermonde determinant implies that [18]:

$$|W| = \prod_{i<l}(\omega_i^2 - \omega_l^2), \quad \Delta_k = \prod_{\substack{i<l \\ i,l \ne k}}(\omega_i^2 - \omega_l^2).$$

By substituting these expressions into (4.54), we obtain the following identity for $m = 1$:

$$|W| = \prod_{i<l}(\omega_i^2 - \omega_l^2) = \sum_{k=1}^{N}(-1)^{k+1}\omega_k^{2(N-1)}\prod_{\substack{i<l \\ i,l \ne k}}(\omega_i^2 - \omega_l^2). \tag{4.55}$$

For an arbitrary fixed j, we also observe that

$$|W| = \prod_{i<l}(\omega_i^2 - \omega_l^2) = \left(\prod_{i<j}(\omega_i^2 - \omega_j^2)\right)\left(\prod_{i>j}(\omega_j^2 - \omega_i^2)\right)\left(\prod_{\substack{i<l \\ i,l \ne j}}(\omega_i^2 - \omega_l^2)\right). \tag{4.56}$$

By taking into account that the rows of (4.54) are linearly dependent for $1 < m \le N$, we get

$$\sum_{k=1}^{N}(-1)^k\omega_k^{2(N-m)}\prod_{\substack{i<l \\ i,l \ne k}}(\omega_i^2 - \omega_l^2) = 0, \quad (1 < m \le N). \tag{4.57}$$

Let us rewrite formula (4.19) for α as follows:

$$\alpha = c_0\omega_1^2 \dots \omega_N^2 \sum_{j=1}^{N}\frac{\omega_j^{2(N-1)}}{\prod_{i \ne j}(\omega_j^2 - \omega_i^2)} = c_0\omega_1^2 \dots \omega_N^2 \sum_{j=1}^{N}\frac{(-1)^{j-1}\omega_j^{2(N-1)}}{\left(\prod_{i<j}(\omega_i^2 - \omega_j^2)\right)\left(\prod_{i>j}(\omega_j^2 - \omega_i^2)\right)}.$$

By using the common denominator $|W|$ for these fractions and applying formula (4.56), we obtain

$$\alpha = \frac{c_0\omega_1^2 \dots \omega_N^2}{|W|}\sum_{j=1}^{N}\left((-1)^{j-1}\omega_j^{2(N-1)}\prod_{\substack{i<l \\ i,l \ne j}}(\omega_i^2 - \omega_l^2)\right).$$

This implies that $\alpha = c_0\omega_1^2\omega_2^2 \dots \omega_N^2$ by taking into account expansion (4.55).

Thus, it remains to prove the expression (4.53) for γ_p. By using the common denominator $|W|$ in (4.28), we obtain from representation (4.56) that

$$
\gamma_p = \frac{1}{|W|} \sum_{k=1}^{N} \left\{ (-1)^{k-1} \omega_k^{2N} \left(\prod_{\substack{i < l \\ i,l \neq k}} (\omega_i^2 - \omega_l^2) \right) \sum_{\substack{1 \leq j_1 < \cdots < j_{N-p} \leq N \\ j_1, \ldots, j_{N-p} \neq k}} (\omega_{j_1} \omega_{j_2} \cdots \omega_{j_{N-p}})^2 \right\}.
$$

(4.58)

By applying the induction, we conclude that

$$
\omega_k^{2N} \sum_{\substack{j_1 < \cdots < j_{N-p} \\ j_1, \ldots, j_{N-p} \neq k}} (\omega_{j_1} \cdots \omega_{j_{N-p}})^2
$$

$$
= \omega_k^{2(N-1)} \sum_{j_1 < \cdots < j_{N-p+1}} (\omega_{j_1} \cdots \omega_{j_{N-p+1}})^2
$$

$$
- \omega_k^{2(N-1)} \sum_{\substack{j_1 < \cdots < j_{N-p+1} \\ j_1, \ldots, j_{N-p+1} \neq k}} (\omega_{j_1} \cdots \omega_{j_{N-p+1}})^2
$$

$$
= \sum_{m=1}^{p} (-1)^{m-1} \omega_k^{2(N-m)} \sum_{j_1 < \cdots < j_{N-p+m}} (\omega_{j_1} \cdots \omega_{j_{N-p+m}})^2.
$$

(4.59)

Let us substitute (4.59) into (4.58) and use property (4.57) for the value of index $m > 1$. As a result, we obtain the terms corresponding to $m = 1$:

$$
\gamma_p = \frac{1}{|W|} \sum_{k=1}^{N} \left\{ (-1)^{k-1} \omega_k^{2(N-1)} \left(\prod_{\substack{i < l \\ i,l \neq k}} (\omega_i^2 - \omega_l^2) \right) \sum_{j_1 < \cdots < j_{N-p+1}} (\omega_{j_1} \cdots \omega_{j_{N-p+1}})^2 \right\}
$$

$$
= \sum_{j_1 < \cdots < j_{N-p+1}} (\omega_{j_1} \cdots \omega_{j_{N-p+1}})^2.
$$

We have used representation (4.55) here. Thus, formulas (4.53) have been proved.

Proof of Theorem 4.5 Consider the optimal control problem for system (4.13) with (4.46) and (4.45) in the class of continuous controls $v(t)$ on $t \in [0, \tau]$. It follows from representation (4.26) and Lemma 4.2 that the optimal control problem considered is equivalent to the Lagrange problem for the minimization of $\tilde{J}(y)$:

$$
\tilde{J}(y) = \int_0^\tau F\left(y, \dot{y}, \ddot{y}, \ldots, y^{(2N+2)}\right) dt \to \min, \tag{4.60}
$$

$$
F\left(y, \dot{y}, \ldots, y^{(2N+2)}\right) = \frac{1}{2} \left\{ \left(\frac{d^2}{dt^2} + \omega_1^2\right) \left(\frac{d^2}{dt^2} + \omega_2^2\right) \cdots \left(\frac{d^2}{dt^2} + \omega_N^2\right) \frac{d^2 y(t)}{dt^2} \right\}^2,
$$

in the class of functions $y \in C^{2N+2}[0, \tau]$ that satisfy boundary conditions (4.51).
Note that a reduction of the optimal control problem for flat systems to the Lagrange
problem in the calculus of variations has been applied, for instance, in the book [23].

Let $y = \bar{y}(t)$ be a solution of the problem (4.60) and (4.51), then the corresponding
control $\bar{v}(t)$, given by formula (4.52), is optimal for the problem (4.13), (4.46), and
(4.45) in the class of continuous controls. If, moreover, $\bar{y} \in C^{4N+4}$ then $\bar{y}(t)$ satisfies
the Euler–Poisson equation:

$$\frac{\partial F}{\partial y} + \sum_{k=1}^{2N+2} (-1)^k \frac{d^k}{dt^k} \left(\frac{\partial F}{\partial y^{(k)}} \right) \Bigg|_{y=\bar{y}(t)} = 0, \quad t \in [0, \tau], \qquad (4.61)$$

and boundary conditions (4.51). Let us introduce a linear differential operator D :
$C^{2N}[0, \tau] \to C[0, \tau]$,

$$y(t) \mapsto Dy(t) = \left(\frac{d^2}{dt^2} + \omega_1^2 \right) \left(\frac{d^2}{dt^2} + \omega_2^2 \right) \cdots \left(\frac{d^2}{dt^2} + \omega_N^2 \right) y(t).$$

Then we can see that Eq. (4.61) can be represented as follows:

$$\frac{d^4}{dt^4} D^2 \bar{y}(t) = 0, \quad t \in [0, \tau]. \qquad (4.62)$$

As solutions $\bar{y}(t)$ of Eq. (4.62) are related to optimal controls $\bar{v}(t)$ by means of
relation (4.52), then $\bar{v}(t) = \frac{1}{c_0 \omega_1^2 \dots \omega_N^2} \frac{d^2}{dt^2} D \bar{y}(t)$ satisfies the following differential
equation:

$$\frac{d^2}{dt^2} D \bar{v}(t) = 0, \quad t \in [0, \tau]. \qquad (4.63)$$

Let us write the characteristic equation for (4.63):

$$\lambda^2 (\lambda^2 + \omega_1^2)(\lambda^2 + \omega_2^2) \cdots (\lambda^2 + \omega_N^2) = 0.$$

This implies that each solution of differential equation (4.63) can be presented in the
form

$$\bar{v}(t) = k_0 + k_1 t + \sum_{j=1}^{N} \left(U_j \cos(\omega_j t) + V_j \sin(\omega_j t) \right), \quad t \in [0, \tau]. \qquad (4.64)$$

According to Theorem 4.3, boundary conditions (4.51) for $y = \bar{y}(t)$ mean that the
corresponding solution $x(t)$ of system (4.13) with control $v = \bar{v}(t)$ has properties
$x(0) = x^0$ and $x(\tau) = x^1$. Instead of the integration of Eq. (4.62) with rather
cumbersome boundary conditions (4.51), let us define the constants k_0, k_1, U_j, V_j

by substituting control (4.64) into the boundary value problem for system (4.13) with $x(0) = x^0$ and $x(\tau) = x^1$. By applying the Cauchy formula for non-homogeneous linear system (4.13) with $v = \bar{v}(t)$ and initial condition $x(0) = x^0$, we get:

$$\begin{pmatrix} \xi_j(t) \\ \eta_j(t) \end{pmatrix} = e^{tA_j} \begin{pmatrix} \xi_j^0 \\ \eta_j^0 \end{pmatrix} + \int_0^t e^{(t-s)A_j} \begin{pmatrix} 0 \\ b_j \bar{v}(s) \end{pmatrix} ds, \quad j = \overline{0, N}, \ (b_0 = 1),$$

(4.65)

where

$$A_0 = \begin{pmatrix} 0 & 1 \\ 0 & 0 \end{pmatrix}, \quad e^{tA_0} = \begin{pmatrix} 1 & t \\ 0 & 1 \end{pmatrix},$$

$$A_j = \begin{pmatrix} 0 & \omega_j \\ -\omega_j & 0 \end{pmatrix}, \quad e^{tA_j} = \begin{pmatrix} \cos \omega_j t & \sin \omega_j t \\ -\sin \omega_j t & \cos \omega_j t \end{pmatrix}, \quad j = \overline{1, N}.$$

By integrating the first two equations of system (4.13) with the control $v = \bar{v}(t)$, we obtain

$$\eta_0(t) = \eta_0^0 + \int_0^t \bar{v}(s) \, ds, \quad \xi_0(t) = \xi_0^0 + \int_0^t \eta_0(s) \, ds.$$

Then we multiply both sides of formula (4.65) with e^{-tA_j} ($1 \le j \le N$) and write the boundary condition $x(\tau) = x^1$ with components $\xi_0, \eta_0, \ldots, \xi_N, \eta_N$. As a result, we obtain system (4.48) with $2N + 2$ linear algebraic equations with respect to $2N + 2$ variables k_0, k_1, U_j, V_j. It remains to show that this system has a unique solution for each $\tau > 0$ and $x^0, x^1 \in \mathbb{R}^{2N+2}$, and that the corresponding control (4.64) is optimal.

For each non-zero control $v \in L^2[0, \tau]$, the functional $J(v)$ is positive in (4.46). Thus, the optimal control problem considered has a solution $\bar{v} \in L^2[0, \tau]$ for each $\tau > 0$ and $x^0, x^1 \in \mathbb{R}^{2N+2}$ by Proposition 16.1 from [24]. In addition, each optimal control $\bar{v}(t)$ is smooth as the maximum of the Hamiltonian is achieved on a linear function of adjoint variables in the Pontryagin maximum principle (cf. [24]). Hence, a smooth solution $y = \bar{y}(t)$ of the Lagrange problem (4.60) exists, and representations (4.26), (4.51) together with formula (4.64) define an optimal control for the coefficients defined by conditions (4.48). If we assume that a solution of system (4.48) is not unique for some $\tau > 0$ and $x^0, x^1 \in \mathbb{R}^{2N+2}$, then it is possible to choose the other boundary conditions $\tilde{x}^0, \tilde{x}^1$ in the right-hand side of (4.48) so that system (4.48) would be inconsistent. This contradicts the existence of optimal controls for system (4.13) for arbitrary boundary conditions $x(0) = \tilde{x}^0$ and $x(\tau) = \tilde{x}^1$. Thus, Theorem 4.5 has been proved.

4.2.4 Approximate Controllability Theorem

The main result of this section is as follows.

Theorem 4.6 *Assume that the parameters of control system (4.11) satisfy the conditions*

$$b_j \neq 0, \quad \omega_j > 0, \quad \omega_i \neq \omega_j \ \text{for all} \ i \neq j,$$

and

$$\sum_{\substack{i, j = 1 \\ i \neq j}}^{\infty} \frac{1}{(\omega_i - \omega_j)^2} < \infty. \tag{4.66}$$

Then there exists a $\tau > 0$ such that such that system (4.11) is approximately controllable in time τ (and the family of controls $u_{x^0, x^1}^N(t)$ defined by formula (4.47) can be used to solve the approximate controllability problem).

Proof With each $N \geq 1$ we associate the operator P_N that projects ℓ^2 onto its $2N + 2$-dimensional subspace spanned by the coordinates $\xi_0, \eta_0, \ldots, \xi_N, \eta_N$:

$$P_N : x = \begin{pmatrix} \xi_0 \\ \eta_0 \\ \vdots \\ \xi_N \\ \eta_N \\ \xi_{N+1} \\ \eta_{N+1} \\ \vdots \end{pmatrix} \in \ell^2 \mapsto P_N x = \begin{pmatrix} \xi_0 \\ \eta_0 \\ \vdots \\ \xi_N \\ \eta_N \\ 0 \\ 0 \\ \vdots \end{pmatrix} \in \ell^2.$$

We introduce the projection operators $Q_N = I - P_N$, where I is the identity operator in ℓ^2. Then the sequence $\{Q_N\}$ satisfy condition (4.3) of Theorem 4.1. For

$$x^i = \begin{pmatrix} \xi_0^i \\ \eta_0^i \\ \xi_1^i \\ \eta_1^i \\ \vdots \end{pmatrix} \in \ell^2, \quad i = 0, 1,$$

and $N \geq 1$, we define the control $u = u_{x^0, x^1}^N(t)$ by formula (4.47). Then condition (4.4) holds by Theorem 4.5.

By taking into account the block structure of A in (4.11), formula (4.2) reads as follows:

$$x(t; x^0, u) = \begin{pmatrix} \xi_0(t) \\ \eta_0(t) \\ \xi_1(t) \\ \eta_1(t) \\ \vdots \end{pmatrix},$$

$$\begin{pmatrix} \xi_j(t) \\ \eta_j(t) \end{pmatrix} = e^{tA_j} \begin{pmatrix} \xi_j^0 \\ \eta_j^0 \end{pmatrix} + b_j \int_0^t e^{(t-s)A_j} \begin{pmatrix} 0 \\ 1 \end{pmatrix} u(s) \, ds, \qquad (4.67)$$

$$b_0 = 1, \quad j = 0, 1, \ldots,$$

where

$$e^{tA_0} = \begin{pmatrix} 1 & t \\ 0 & 1 \end{pmatrix}, \quad e^{tA_j} = \begin{pmatrix} \cos \omega_j t & \sin \omega_j t \\ -\sin \omega_j t & \cos \omega_j t \end{pmatrix}, \quad j \geq 1. \qquad (4.68)$$

In order to check condition (4.5), we estimate the L^2-norm of control functions $u^N = u_{x^0, x^1}^N(t)$:

$$\|u^N\|_{L^2(0,\tau)}^2 = \int_0^\tau (u^N(t))^2 \, dt$$

$$= \int_0^\tau \{k_0 + k_1 t + \sum_{j=1}^N (U_j \cos(\omega_j t) + V_j \sin(\omega_j t))\} u^N dt. \qquad (4.69)$$

For computing (4.69), we multiply formula (4.67) with $e^{-\tau A_j}$, $j = \overline{0, N}$:

$$e^{-\tau A_j} \begin{pmatrix} \xi_j^1 \\ \eta_j^1 \end{pmatrix} - \begin{pmatrix} \xi_j^0 \\ \eta_j^0 \end{pmatrix} = b_j \int_0^\tau e^{-t A_j} \begin{pmatrix} 0 \\ 1 \end{pmatrix} u^N(t) \, dt.$$

This implies, taking into account (4.68), that

$$\begin{pmatrix} \xi_0^1 - \xi_0^0 - \tau \eta_0^1 \\ \eta_0^1 - \eta_0^0 \end{pmatrix} = \int_0^\tau \begin{pmatrix} -t u^N(t) \\ u^N(t) \end{pmatrix} dt,$$

$$\begin{pmatrix} \xi_j^1 \cos \omega_j \tau - \eta_j^1 \sin \omega_j \tau - \xi_j^0 \\ \xi_j^1 \sin \omega_j \tau + \eta_j^1 \cos \omega_j \tau - \eta_j^0 \end{pmatrix} = b_j \int_0^\tau \begin{pmatrix} -\sin(\omega_j t) u^N(t) \\ \cos(\omega_j t) u^N(t) \end{pmatrix} dt, \quad j = \overline{1, N}.$$

By substituting these integrals into (4.69), we get

$$\|u^N\|^2_{L^2(0,\tau)} = (\eta_0^1 - \eta_0^0)k_0 + \left(\frac{\xi_0^0 - \xi_0^1}{\tau} + \eta_0^1\right)\tau k_1$$

$$+ \sum_{j=1}^{N}(\xi_j^1 \sin\omega_j\tau + \eta_j^1 \cos\omega_j\tau - \eta_j^0)\frac{U_j}{b_j}$$

$$- \sum_{j=1}^{N}(\xi_j^1 \cos\omega_j\tau - \eta_j^1 \sin\omega_j\tau - \xi_j^0)\frac{V_j}{b_j}. \tag{4.70}$$

It is easy to see that the coefficients at U_j, V_j in formula (4.70) coincide (up to their sign) with the expressions in the right-hand side of (4.48). For system (4.48), let us denote the vector of unknowns by $\varkappa = (k_0, \tau k_1, U_1, V_1, \ldots, U_N, V_N)^T$ and the right-hand side by y. Then system (4.48) takes the form $(M + F)\varkappa = y$, or, equivalently,

$$\varkappa = M^{-1}y - M^{-1}F\varkappa,$$

as the matrix M is non-degenerate for all $\tau > 0$. This implies that

$$\|\varkappa\| \le \|M^{-1}y\| + \|M^{-1}\| \cdot \|F\| \cdot \|\varkappa\|.$$

If $\|M^{-1}\| \cdot \|F\| < 1$ then

$$\|\varkappa\| \le \frac{\|M^{-1}y\|}{1 - \|M^{-1}\| \cdot \|F\|}. \tag{4.71}$$

By computing the inverse matrix for (4.49), we get

$$\frac{\tau}{2} \cdot M^{-1} = \begin{pmatrix} \begin{pmatrix} 3 & -1 \\ -6 & 3 \end{pmatrix} & O & \cdots & O \\ O & \begin{pmatrix} 1 & 0 \\ 0 & -1 \end{pmatrix} & \cdots & O \\ \vdots & \vdots & \ddots & \vdots \\ O & O & \cdots & \begin{pmatrix} 1 & 0 \\ 0 & -1 \end{pmatrix} \end{pmatrix}.$$

Then straightforward computations show that

$$\|M^{-1}\| = \sup_{\|y\|=1} \|M^{-1}y\| = \frac{\sqrt{61} + 7}{\tau},$$

where $\|y\|$ stands for the Euclidean norm of a vector y in $\mathbb{R}^{2N+2}$.

Let us estimate the norm of matrix F:

$$\|F\|^2 \leq \sum_{l,k=1}^{N} f_{lk}^2 \leq \sum_{l,k=1}^{\infty} f_{lk}^2, \tag{4.72}$$

where f_{lk} are given by (4.50). Condition (4.66) implies that the sequence of positive numbers $\{\omega_j\}_{j=1}^{\infty}$ has the minimal element $\omega_{min} > 0$. Hence, the following inequalities hold:

$$\sum_{j=1}^{\infty} \frac{1}{\omega_j^2} < \frac{1}{\omega_{min}^2} + \sum_{\substack{i,j=1 \\ i \neq j}}^{\infty} \frac{1}{(\omega_i - \omega_j)^2} < \infty,$$

$$\sum_{\substack{i,j=1 \\ i \neq j}}^{\infty} \frac{1}{(\omega_i + \omega_j)^2} < \sum_{\substack{i,j=1 \\ i \neq j}}^{\infty} \frac{1}{(\omega_i - \omega_j)^2} < \infty. \tag{4.73}$$

Let us fix a number $\tau_0 > 0$. If follows from (4.72), (4.50), and (4.73) that

$$\|F\|^2 = O\left(\sum_{\substack{i,j=1 \\ i \neq j}}^{\infty} \frac{1}{(\omega_i - \omega_j)^2} \right)$$

uniformly on $\tau \geq \tau_0$. Hence, there is a positive constant $\bar{H} < \infty$ such that

$$\|F\| \leq \bar{H}, \quad \forall N \geq 1, \quad \forall \tau \geq \tau_0. \tag{4.74}$$

Let us choose a τ from the conditions

$$\tau \geq \tau_0, \quad \tau > (\sqrt{61} + 7)\bar{H}.$$

Note that such a τ does not depend on the boundary conditions x^0 and x^1. Estimate (4.71) yields

$$\|\varkappa\| \leq \frac{\tau \|M^{-1} y\|}{\tau - (\sqrt{61} + 7)\bar{H}} = \frac{\tau}{\tau - (\sqrt{61} + 7)\bar{H}}$$

$$\times \left\{ \left(\frac{\xi_0^1 - \xi_0^0}{\tau} - \eta_0^0 \right)^2 + \left(\eta_0^1 - \eta_0^0 \right)^2 + \delta_N^2 \right\}^{1/2}, \tag{4.75}$$

where

$$\delta_N^2 = \sum_{j=1}^{N} \frac{(\xi_j^1 \sin \omega_j \tau + \eta_j^1 \cos \omega_j \tau - \eta_j^0)^2}{b_j^2}$$

$$+ \sum_{j=1}^{N} \frac{(\xi_j^1 \cos \omega_j \tau - \eta_j^1 \sin \omega_j \tau - \xi_j^0)^2}{b_j^2}$$

$$\leq 2 \sum_{j=1}^{N} \left(\frac{(\xi_j^0)^2 + (\eta_j^0)^2 + (\xi_j^1)^2 + (\eta_j^1)^2}{b_j^2} \right). \qquad (4.76)$$

We have used the inequality $(a+b)^2 \leq 2(a^2+b^2)$ here. By applying the Cauchy–Schwartz inequality in (4.70), we obtain

$$\|u^N\|_{L^2}^2 \leq \|\varkappa\| \left\{ (\eta_0^1 - \eta_0^0)^2 + \left(\frac{\xi_0^0 - \xi_0^1}{\tau} + \eta_0^1 \right)^2 + \delta_N^2 \right\}^{1/2}.$$

By using the inequality for $\|\varkappa\|$ from (4.75) and estimate for δ_N^2 from (4.76), we get

$$\|u^N\|_{L^2(0,\tau)}^2 \leq \hat{H} \left\{ (\xi_0^0)^2 + (\eta_0^0)^2 + (\xi_0^1)^2 + (\eta_0^1)^2 \right.$$

$$\left. + \sum_{j=1}^{N} \left(\frac{(\xi_j^0)^2 + (\eta_j^0)^2 + (\xi_j^1)^2 + (\eta_j^1)^2}{b_j^2} \right) \right\} \qquad (4.77)$$

with some positive constant $\hat{H}$. Consider the following dense subset of ℓ^2,

$$S = \{x \in \ell^2 \mid \lim_{n \to \infty} \left(\sum_{j=n+1}^{\infty} b_j^2 \right) \left(\sum_{j=1}^{n} \frac{\xi_j^2 + \eta_j^2}{b_j^2} \right) = 0\}.$$

As $B \in \ell^2$ then

$$\|Q_N B\|_{\ell^2}^2 = \sum_{j=N+1}^{\infty} b_j^2 \to 0 \quad \text{as } N \to \infty.$$

For any $x^0, x^1 \in S$, estimate (4.77) implies that

$$\lim_{N \to \infty} \left\{ \left(\sum_{j=N+1}^{\infty} b_j^2 \right) \left((\xi_0^0)^2 + (\eta_0^0)^2 + (\xi_0^1)^2 + (\eta_0^1)^2 \right. \right.$$

$$\left. \left. + \sum_{j=1}^{N} \left(\frac{(\xi_j^0)^2 + (\eta_j^0)^2 + (\xi_j^1)^2 + (\eta_j^1)^2}{b_j^2} \right) \right) \right\} = 0,$$

so condition (4.5) holds. Therefore, system (4.11) is approximately controllable by Theorem 4.1.

Remark 4.1 An open question is whether it is possible to relax conditions (4.66) in order to justify the relevance of controls (4.10) under a weaker assumption on the distribution of the modal frequencies $\{\omega_n\}$.

4.3 Application for the Euler–Bernoulli Beam

In order to justify the applicability of Theorem 4.6, we check the convergence of series in (4.66) for a model of a rotating rigid body with attached Euler–Bernoulli beams. For simplicity, we consider a mechanical system with one flexible beam.

Such system is described by differential equation (4.11) with coefficients given by formulas (4.12), i.e.

$$\omega_n = c_1\sqrt{\lambda_{1n}}, \ b_n = -J_{1n}/\rho_1, \quad n = 1, 2, \dots \qquad (4.78)$$

where λ_{1n} are eigenvalues of the spectral problem (3.10) and (3.11), and J_{1n} are defined by relation (3.19).

We use representations (4.78) to verify condition (4.66).

Lemma 4.3 *Let ω_n be defined by formula (4.78) for $n = 1, 2, \dots$ Then*

$$\sum_{\substack{i, j = 1 \\ i \neq j}}^{\infty} \frac{1}{(\omega_i - \omega_j)^2} < \infty. \qquad (4.79)$$

Proof Let us use representation (3.15) for the eigenvalues:

$$\lambda_{1j} = (\beta_j/l_1)^4,$$

where $l_1 = l$ is the beam length and $\beta_1 < \beta_2 < \cdots$ are positive roots of Eq. (3.13),

$$1 + \cos(\beta_j)\cosh\left(\beta_j\right) = 0.$$

Let us rewrite this equation in the form

$$\cos(\beta_j) = \varphi(\beta_j), \quad \varphi(\beta_j) = -1/\cosh(\beta_j) \qquad (4.80)$$

Since $\varphi(\beta) \to 0$ and $\varphi'(\beta) \to 0$ as $\beta \to +\infty$, then solutions of Eq. (4.80) can be approximated by zeroes of the function $\cos(\beta_j)$ for "large" β_j, i.e. for each $\delta > 0$, there is an $n_0 = n_0(\delta) > 1$ such that

$$\omega_-(j) < \omega_j < \omega_+(j) \tag{4.81}$$

for all $j \geq n_0$. Here

$$\omega_-(j) = \frac{\pi^2 c}{l^2}\left(j - \frac{1}{2} - \delta\right)^2, \quad \omega_+(j) = \frac{\pi^2 c}{l^2}\left(j - \frac{1}{2} + \delta\right)^2. \tag{4.82}$$

Let us choose a positive number δ such that the following conditions hold

$$\omega_-(n_0) > \omega_{n_0-1}, \quad \omega_-(j+1) > \omega_+(j)$$

for all $j \geq n_0(\delta)$. It can be shown that these conditions are satisfied for $\delta < 1/2$. As all terms of the series (4.79) are positive, then a permutation of terms does not affect the convergence of this series. Let us rewrite series (4.79) in the form

$$\frac{1}{2}\sum_{\substack{i,j=1 \\ i \neq j}}^{\infty} \frac{1}{(\omega_i - \omega_j)^2} = \sum_{\substack{j \geq 1 \\ i > j}} \frac{1}{(\omega_i - \omega_j)^2} = \sum_{\substack{1 \leq j < n_0 \\ j < i < n_0}} \frac{1}{(\omega_i - \omega_j)^2}$$

$$+ \sum_{\substack{1 \leq j < n_0 \\ i \geq n_0}} \frac{1}{(\omega_i - \omega_j)^2} + \sum_{\substack{j \geq n_0 \\ i > j}} \frac{1}{(\omega_i - \omega_j)^2}.$$

Note that the first sum in the brackets is finite. Thus, for the convergence of (4.79) it suffices to show that

$$\sum_{\substack{1 \leq j < n_0 \\ i \geq n_0}} \frac{1}{(\omega_i - \omega_j)^2} < \infty, \quad \sum_{\substack{j \geq n_0 \\ i > j}} \frac{1}{(\omega_i - \omega_j)^2} < \infty. \tag{4.83}$$

In order to study series (4.83) we apply an inequality used in the integral convergence test. Namely, if $f : [n_0, \infty) \to \mathbb{R}$ is a non-increasing function with non-negative values, then

$$\sum_{j=n_0}^{\infty} f(j) \leq f(n_0) + \int_{n_0}^{\infty} f(y)dy. \tag{4.84}$$

Now we apply inequalities (4.81) and (4.84) to series (4.83):

$$\sum_{\substack{1 \le j < n_0 \\ i \ge n_0}} \frac{1}{(\omega_i - \omega_j)^2} < \sum_{j=1}^{n_0-1} \sum_{i=n_0}^{\infty} \frac{1}{(\omega_-(i) - \omega_j)^2}$$

$$\le \sum_{j=1}^{n_0-1} \left\{ \frac{1}{(\omega_-(n_0) - \omega_j)^2} + \int_{n_0}^{\infty} \frac{dy}{(\omega_-(y) - \omega_j)^2} \right\}, \quad (4.85)$$

$$\sum_{\substack{j \ge n_0 \\ i > j}} \frac{1}{(\omega_i - \omega_j)^2} < \sum_{\substack{j \ge n_0 \\ i > j}} \frac{1}{(\omega_-(i) - \omega_+(j))^2}$$

$$\le \sum_{j=n_0}^{\infty} \left\{ \frac{1}{(\omega_-(j+1) - \omega_+(j))^2} + \int_{j+1}^{\infty} \frac{dy}{(\omega_-(y) - \omega_+(j))^2} \right\}.$$
$$(4.86)$$

As the quadratic function $\omega_-(y)$ is defined in (4.82), and $\omega_-(y) - \omega_j \ge \omega_-(n_0) - \omega_j > 0$ for all $y \ge n_0$ and $j \le n_0 - 1$ with the chosen δ, then the integrand of (4.85) admits the following estimate:

$$\frac{1}{(\omega_-(y) - \omega_j)^2} = O(y^{-4})$$

for $y \ge n_0$ and $j \le n_0 - 1$.

This implies that the integral in the right-hand side of (4.85) converges for each index $j \le n_0 - 1$. Hence, the series in the left-hand side of (4.85) converges. In order to estimate the right-hand side of formula (4.86), we apply representation (4.82) and inequality (4.84):

$$\sum_{j=n_0}^{\infty} \frac{1}{(\omega_-(j+1) - \omega_+(j))^2} = \frac{l^4}{\pi^4 c^2} \sum_{j=n_0}^{\infty} \left\{ \left(j + \frac{1}{2} - \delta \right)^2 + \left(j - \frac{1}{2} + \delta \right)^2 \right\}^{-2}$$

$$= \frac{l^4}{4\pi^4 c^2 (1 - 2\delta)^2} \sum_{j=n_0}^{\infty} j^{-2},$$

$$\sum_{j=n_0}^{\infty} \int_{j+1}^{\infty} \frac{dy}{(\omega_-(y) - \omega_+(j))^2} = \frac{l^4}{\pi^4 c^2} \sum_{j=n_0}^{\infty} \int_{j+1}^{\infty} \left\{ \left(y - \frac{1}{2} - \delta \right)^2 + \left(j - \frac{1}{2} + \delta \right)^2 \right\}^{-2} dy$$

$$= \frac{l^4}{\pi^4 c^2} \sum_{j=n_0}^{\infty} \left\{ \frac{1}{(2j - 1 + 2\delta)^3} \left(2 \ln \left(\frac{y + j - 1}{y - j - 2\delta} \right) \right. \right.$$

$$\left. \left. - \frac{(2y - 1 - 2\delta)(2j - 1 - 2\delta)}{(y - j - 2\delta)(y + j - 1)} \right) \right\} \Bigg|_{y=j+1}^{\infty}$$

$$= \frac{l^4}{\pi^4 c^2} \sum_{j=n_0}^{\infty} \left\{ \frac{1}{(2j-1+2\delta)^3} \left(2\ln\left(\frac{1-2\delta}{2j} \right) \right. \right.$$
$$\left. \left. + \frac{(2j+1-2\delta)(2j-1+2\delta)}{2j(1-2\delta)} \right) \right\}$$

$$= \sum_{j=n_0}^{\infty} O(j^{-2}).$$

Thus, the right-hand side of inequality (4.86) is dominated by a convergent series, which proves the assertion of this Lemma.

By exploiting formulas (3.13) and (3.15) we conclude that condition (4.66) holds if $\{\omega_n\}$ are given by representation (4.12). Thus, Theorem 4.6 implies the following corollary.

Corollary of Theorem 4.6 *Let the coefficients of control system* (4.11) *be defined by formulas* (4.12), (3.15). *Then system* (4.11) *is approximately controllable in some time* τ. *Moreover, the family of controls* $u_{x^0,x^1}^N(t)$, *given by* (4.47), *solves the approximate controllability problem.*

Proof Lemma 4.3 implies that the coefficients ω_j satisfy condition (4.66). From representations (4.78) and (3.19) it follows that $b_j \neq 0$ for all $j = 1, 2, \ldots$ It is easy to see that

$$S_0 = \{x \in \ell^2 | \, \exists n_0 : \quad \xi_j = \eta_j = 0, \quad \forall j \geq n_0\},$$

and the set S_0 is dense in ℓ^2.

References

1. Balakrishnan, A.V.: Applied Functional Analysis, 2nd edn. Springer, New York (1981)
2. Coron, J.-M.: Control and Nonlinearity. AMS, Providence (2007)
3. Krabs, W.: On Moment Theory and Controllability of One Dimensional Vibrating Systems and Heating Processes. Lecture Notes in Control and Information Sciences, vol. 173. Springer, Berlin (1992)
4. Lagnese, J.E., Leugering, G.: Controllability of thin elastic beams and plates. In: Levine, W.S. (ed.) The Control Handbook, pp. 1139–1156. CRC Press—IEEE Press, Boca Raton (1996)
5. Lagnese, J., Lions, J.L.: Modelling Analysis and Control of Thin Plates. Springer, Masson (1988)
6. Levan, N., Rigby, L.: Strong stabilizability of linear contractive control systems on Hilbert space. SIAM J. Control Optim. **17**, 23–35 (1979)
7. Tucsnak, M., Weiss, G.: Observation and Control for Operator Semigroups. Birkhäuser, Basel (2009)
8. Chernousko, F.L., Ananievski, I.M., Reshmin, S.A.: Control of Nonlinear Dynamical Systems. Methods and Applications. Springer, Berlin (2008)
9. Dáger, R., Zuazua, E.: Wave Progagation, Observation and Control in 1-d Flexible Multi-Structures. Springer, Berlin (2006)

10. Jacob, B., Partington, J.R.: On controllability of diagonal systems with one-dimensional input space. Syst. Control Lett. **55**, 321–328 (2006)
11. Fattorini, H.O.: Infinite Dimensional Optimization and Control Theory. Cambridge University Press, Cambridge (1999)
12. Balas, M.J.: Modal control of certain flexible dynamical systems. SIAM J. Control Optim. **16**, 450–462 (1978)
13. Pazy, A.: Semigroups of Linear Operators and Applications to Partial Differential Equations. Springer, New York (1983)
14. Brunovský, P.: A classification of linear controllable systems. Kybernetica. **6**, 173–188 (1970)
15. Fliess, M., Lévine, J., Martin, Ph, Rouchon, P.: Flatness and defect of nonlinear systems: introductory theory and examples. Int. J. Control. **61**(6), 1327–1361 (1995)
16. Rudolph, J.: Flatness Based Control of Distributed Parameter Systems. Shaker Verlag, Aachen (2003)
17. Rouchon, P.: Flatness based control of oscillators. ZAMM Z. Angew. Math. Mech. **85**, 411–421 (2005)
18. Gantmacher, F.R.: The Theory of Matrices. AMS, Providence (1959)
19. Fliess, M., Lévine, J., Martin, Ph, Rouchon, P.: Open problems in mathematical systems and control theory. In: Blondel, V., et al. (eds.) Some Open Questions Related to Flat Nonlinear Systems, pp. 99–103. Springer, London (1999)
20. Charlet, B., Lévine, J., Marino, R.: On dynamic feedback linearization. Syst. Control Lett. **13**, 143–151 (1989)
21. Jakubczyk, B., Respondek, W.: On linearization of control systems. Bull. Acad. Polonaise Sci., Ser. Sci. Math. **28**, 517–522 (1980)
22. Isidori, A.: Nonlinear Control Systems, 3rd edn. Springer, New York (1995)
23. Sira-Ramírez, H., Agrawal, S.K.: Differentially Flat Systems. Taylor & Francis, UK (2004)
24. Agrachev, A.A., Sachkov, Yu. L.: Control Theory from the Geometric Viewpoint. Springer, Berlin (2004)

Chapter 5
Observer-Based Stabilization of a Manipulator Based on the Timoshenko Beam Model

Abstract A mathematical model of a flexible-link manipulator is studied in this chapter within the framework of the Timoshenko beam theory. In contrast to the majority of publications in this area, we consider here the model with two rigid bodies and the beam with non-collocated sensors and actuators. For the mathematical model described by coupled ordinary and partial differential equations, we construct a family of Galerkin's approximations with an arbitrary number of modal coordinates. Asymptotic properties of the eigenvalues of the associated spectral problem are investigated for the case of a homogeneous beam. We propose a state feedback law and an observer-based stabilization scheme for Galerkin's systems.

An extension of the Euler–Bernoulli model was proposed by Timoshenko [1]. From the mechanical viewpoint, the Timoshenko beam has an advantage of describing the effects of rotary inertia and the deflection due to shear. Issues of modeling and control of Timoshenko beams have been studied, for instance, in the monograph [2] and papers [3–15]. Networks of the Euler–Bernoulli and Timoshenko beams have been considered in the monograph [16]. Spectral analysis of coupled Euler–Bernoulli and Timoshenko beams is carried out in the paper [17]. The motion of a payload, usually attached to a real manipulator, is neglected in all these publications. In paper [18], a clamped beam with an end mass is proved to be stabilizable by a feedback control applied to the tip. The author of [19] addresses the development of LQR techniques and computation algorithms for beams with control torques applied to the hub. A limitation of these results is that a knowledge of the full infinite-dimensional space is required. It should be emphasized that, in contrast to the above publications, we study here a rotating beam that carries a payload under the action of gravity, the control torque is applied at the hub, and the shear motion is taken into account. The motivation for this study is to control the motion of a real flexible-link manipulator—turntable ladder.

5.1 Description of the Model

Figure 5.1 shows the scheme of a manipulator that consists of two rigid bodies B_0 (hub) and B_1 (payload) interconnected by an elastic beam of the length l.

© Springer International Publishing Switzerland 2015   131
A.L. Zuyev, *Partial Stabilization and Control of Distributed Parameter Systems with Elastic Elements*, Lecture Notes in Control and Information Sciences 458,
DOI 10.1007/978-3-319-11532-0_5

Fig. 5.1 Flexible-link
manipulator with the
Timoshenko beam

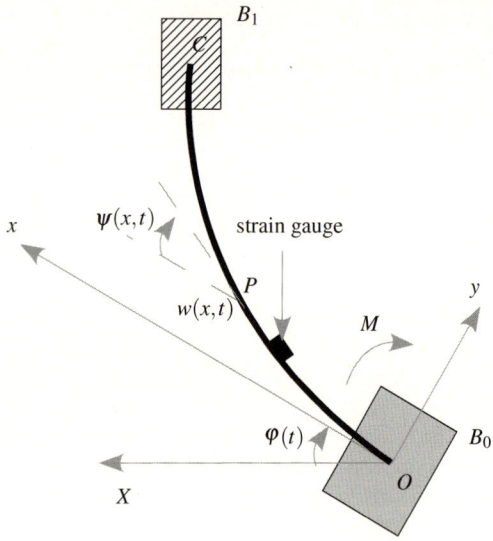

This mechanical system rotates in the vertical plane around the fixed point O under the action of the control torque M applied at the hub B_0. Let us denote by Oxy the Cartesian frame associated with B_0. The orientation of B_0 at time t is defined by the angle $\varphi(t)$ between the fixed horizontal axis OX and Ox. We assume that one end of the beam is clamped at the point O of B_0, while the other end is clamped at the center of mass C of B_1. We also assume that the centerline of the beam in its undeformed configuration occupies the segment $[0, l]$ on the Ox-axis. Our assumptions are based on the manipulator model described in the paper [20].

Consider a point P of the beam with the Lagrangian coordinate x, $0 \leq x \leq l$. Then the coordinates of P in the frame Oxy are $(x + s(x, t), w(x, t))$, where functions $w(x, t)$ and $s(x, t)$ describe the transverse and longitudinal components of the displacement at time t, respectively. Following the Timoshenko beam theory [1], we introduce the angle $\psi(x, t)$ to describe the rotation of the cross-section of the beam about the direction of Ox. Thus, the configuration of the manipulator at time $t \geq 0$ is defined by functions $\varphi(t), w(x, t), \psi(x, t), 0 \leq x \leq l$.

Under assumptions of the Timoshenko beam theory, the kinetic energy of the system takes the following form:

$$
\begin{aligned}
2T = \int_0^l & \{\rho(x)[(\dot{w} + x\dot{\varphi})^2 + (w\dot{\varphi})^2 + \dot{s}^2 + 2\dot{\varphi}(\dot{w}s - \dot{s}w) + (\dot{\varphi})^2(s + 2x)s] \\
& + I_\rho(x)(\dot{\varphi} + \dot{\psi})^2\}dx + J_0(\dot{\varphi})^2 + m\{(\dot{w} + x\dot{\varphi})^2 + (w\dot{\varphi})^2 + \dot{s}^2 \\
& + 2\dot{\varphi}(\dot{w}s - \dot{s}w) + (\dot{\varphi})^2(s + 2x)s\}\big|_{x=l} + J_c\{\dot{\varphi} + \dot{\psi}\}^2\big|_{x=l},
\end{aligned} \tag{5.1}
$$

where $\rho(x)$ is the mass per unit length of the beam (linear density), $I_\rho(x)$ is the mass moment of inertia of the cross-section, J_0 is the moment of inertia of B_0 with respect to the point O, m is the mass of B_1, and J_c is the moment of inertia of B_1 with respect to its center of mass C. Here, and in the sequel, the dot denotes the differentiation with respect to time t, and the prime stands for the differentiation with respect to the space variable x.

Assuming that the beam is inextensible, we get the following relation for small deformations:

$$s' = -\frac{1}{2}w'^2 + o(w'^2).$$

The integration of this relation implies the following formula (up to the higher order terms):

$$s(x, t) = -\frac{1}{2}\int_0^x w'^2(\xi, t)\, d\xi. \tag{5.2}$$

To derive the equations of motion, we assume that the deformation of the beam is small and neglect the terms of order 2 and higher with respect to w and its derivatives.

Following the Timoshenko beam theory [1] and using formulas (5.1) and (5.2), we get the Lagrangian of the system considered (cf. [20]):

$$
\begin{aligned}
2L = &\int_0^l \left\{ \rho(x)\left((\dot{w} + x\dot{\varphi})^2 + \dot{\varphi}^2 w^2 \right) - \rho_2(x)\dot{\varphi}^2 w'^2 + I_\rho(\dot{\varphi} + \dot{\psi})^2 \right. \\
&\left. - K(\psi - w')^2 - EI(\psi')^2 \right\} dx + m\left\{ \dot{\varphi}^2 w^2(l, t) + (l\dot{\varphi} + \dot{w}(l, t))^2 \right\} \\
&+ J_c\left\{ \dot{\varphi} + \dot{\psi}(l, t) \right\}^2 + J_0\dot{\varphi}^2 - g\int_0^l \left\{ (2\rho x - \rho_1 w'^2)\sin\varphi + 2\rho w\cos\varphi \right\} dx \\
&- 2mg\left\{ l\sin\varphi + w(l, t)\cos\varphi \right\},
\end{aligned}
\tag{5.3}
$$

where

$$\rho_1(x) = \int_x^l \rho(\xi)\, d\xi + m, \quad \rho_2(x) = \int_x^l \xi\rho(\xi)\, d\xi + ml.$$

Here E is the Young's modulus, I is the moment of inertia of the cross-section of the beam, and g is the gravity constant. The coefficient K is equal to γGA, where G is the modulus of elasticity in shear, A is the cross-section area, and the constant γ characterizes the geometry of the beam section. We assume that ρ, I_ρ, EI, K are positive differentiable functions of the space variable x.

If the motion of the mechanical system is defined by C^2-functions

$$(\varphi(t), w(x, t), \psi(x, t))$$

under the action of the control torque $M(t)$, for $t \in [t_1, t_2]$, then Hamilton's principle implies (see [21]):

$$\delta \left(\int_{t_1}^{t_2} L \, dt \right) + \int_{t_1}^{t_2} M(t) \delta \varphi(t) \, dt = 0, \tag{5.4}$$

for any admissible variations $(\delta \varphi(t), \delta w(x, t), \delta \psi(x, t))$ satisfying the boundary conditions

$$\delta \varphi|_{t=t_1} = \delta \varphi|_{t=t_2} = 0, \quad \delta w|_{t=t_1} = \delta w|_{t=t_2} = 0, \quad \delta \psi|_{t=t_1} = \delta \psi|_{t=t_2} = 0,$$

$$\delta w|_{x=0} = 0, \quad \delta \psi|_{x=0} = 0. \tag{5.5}$$

The computation of the first variation in (5.4) and the integration by parts yield

$$\int_{t_1}^{t_2} \left\{ \left(M + \frac{\partial L}{\partial \varphi} - \frac{d}{dt} \frac{\partial L}{\partial \dot{\varphi}} \right) \delta \varphi(t) - \mu \left(\delta w(\cdot, t), \delta \psi(\cdot, t); \varphi, w, \psi \right) \right\} dt = 0,$$

$$\tag{5.6}$$

where μ is a linear functional with respect to δw and $\delta \psi$:

$$\mu \left(\delta w(\cdot, t), \delta \psi(\cdot, t); \varphi, w, \psi \right) = \int_0^l \delta w(x, t) \left\{ (\ddot{w} + x\ddot{\varphi} - (\dot{\varphi})^2 w + g \cos \varphi) \rho \right.$$

$$\left. + \left(K(\psi - w') + (g\rho_1 \sin \varphi - (\dot{\varphi})^2 \rho_2) w' \right)' \right\} dx$$

$$+ \int_0^l \delta \psi(x, t) \left\{ I_\rho(\ddot{\psi} + \ddot{\varphi}) + K(\psi - w') - (EI\psi')' \right\} dx$$

$$+ \delta \psi(l, t) \left\{ J_c(\ddot{\varphi} + \ddot{\psi}) + EI\psi' \right\} \big|_{x=l}$$

$$+ \delta w(l, t) \left\{ K(w' - \psi) + m(\ddot{w} + l\ddot{\varphi} - (\dot{\varphi})^2 w + g \cos \varphi) \right.$$

$$\left. + m(l(\dot{\varphi})^2 - g \sin \varphi) w' \right\} \big|_{x=l}. \tag{5.7}$$

Since the expression (5.6) is equal to zero for any admissible variations satisfying conditions (5.5), we deduce the following manipulator motion equations from (5.6) (cf. [20]):

$$\ddot{w} + \frac{1}{\rho}\left(K(\psi - w')\right)' = -g\cos\varphi - x\ddot{\varphi} + \dot{\varphi}^2 w + \frac{1}{\rho}\left((\rho_2\dot{\varphi}^2 - g\rho_1\sin\varphi)w'\right)';$$

$$\ddot{\psi} + \frac{K}{I_\rho}(\psi - w') - \frac{1}{I_\rho}(EI\psi')' = -\ddot{\varphi}, \quad x \in (0, l);$$

$$w|_{x=0} = \psi|_{x=0} = 0;$$

$$K(\psi - w')\big|_{x=l} = m\left\{\ddot{w} + l\ddot{\varphi} - \dot{\varphi}^2 w + g\cos\varphi + (l\dot{\varphi}^2 - g\sin\varphi)w'\right\}\big|_{x=l};$$

$$-EI\psi'\big|_{x=l} = J_c(\ddot{\varphi} + \ddot{\psi}\big|_{x=l}),$$

$$M(t) = \frac{d}{dt}\frac{\partial L}{\partial\dot{\varphi}} - \frac{\partial L}{\partial\varphi}$$

$$= \left\{J_c + J_0 + m[l^2 + w^2(l, t)] + \int_0^l [I_\rho + (x^2 + w^2)\rho - \rho_2 w'^2]\,dx\right\}\ddot{\varphi}$$

$$+ \int_0^l \left(\rho x\ddot{w} + I_\rho\ddot{\psi} + 2\rho w\dot{\varphi}\dot{w} - 2\rho_2 w'\dot{\varphi}\dot{w}'\right)dx$$

$$+ m\,(l\ddot{w} + 2w\dot{\varphi}\dot{w})|_{x=l} + J_c\,\ddot{\psi}\big|_{x=l}$$

$$+ g\left\{\int_0^l \left(\rho x - \frac{1}{2}\rho_1 w'^2\right)dx + ml\right\}\cos\varphi$$

$$- g\left\{\int_0^l \rho w\,dx + m\,w(l, t)\right\}\sin\varphi. \tag{5.8}$$

By direct substitution we see that the boundary value problem (5.8) admits a solution

$$\varphi(t) = \varphi_0, \quad w(x, t) = w_0(x), \quad \psi(x, t) = \psi_0(x), \quad M(t) = M_0 \tag{5.9}$$

provided that

$$\left(K(w_0'(x) - \psi_0(x))\right)' = g\left\{\rho\cos\varphi_0 + (\rho_1 w_0')'\sin\varphi_0\right\};$$

$$(EI\psi_0'(x))' + K(w_0'(x) - \psi_0(x)) = 0, \quad x \in (0, l);$$

$$w_0(0) = \psi_0(0) = 0; \quad \psi_0'(l) = 0; \quad K(\psi_0(l) - w_0'(l)) = mg\left(\cos\varphi_0 - w_0'(l)\sin\varphi_0\right);$$

$$\frac{M_0}{g} = \left(\int_0^l \left(\rho x - \frac{1}{2}\rho_1 w_0'^2 \right) dx + ml \right) \cos \varphi_0 - \left(\int_0^l \rho w_0 \, dx + m\, w_0(l) \right) \sin \varphi_0.$$

(5.10)

Solution (5.9) corresponds to an equilibrium of the manipulator considered. Our next goal is to construct a finite dimensional reduced system for the boundary value problem (5.8) in a neighborhood of the equilibrium (5.9).

5.2 Equations of the Perturbed Motion

Let (φ_0, w_0, ψ_0) be a solution of system (5.10) for some torque M_0. By putting

$$\varphi = \varphi_0 + \tilde{\varphi}, \; w = w_0 + \tilde{w}, \; \psi = \psi_0 + \tilde{\psi}, \; M = M_0 + \tilde{M}$$

into dynamical equations (5.8), we get the following boundary value problem:

$$\ddot{\tilde{\varphi}} = v;$$

$$\ddot{\tilde{w}} + \frac{1}{\rho}\left(K(\tilde{\psi} - \tilde{w}')\right)' = -xv + g\tilde{\varphi}\sin\varphi_0 - \frac{g}{\rho}\left((\tilde{w}'\sin\varphi_0 + \tilde{\varphi}\, w_0'\cos\varphi_0)\rho_1\right)' + \cdots;$$

$$I_\rho \ddot{\tilde{\psi}} + K(\tilde{\psi} - \tilde{w}') - \left(EI\tilde{\psi}'\right)' = -I_\rho v;$$

$$\tilde{w}|_{x=0} = \tilde{\psi}\Big|_{x=0} = 0;$$

$$\left(\frac{K}{m}(\tilde{w}' - \tilde{\psi}) + \ddot{\tilde{w}} - g(\sin\varphi_0 + w_0'\cos\varphi_0)\tilde{\varphi} - g\tilde{w}'\sin\varphi_0 + \cdots\right)\Bigg|_{x=l} = -lv;$$

(5.11)

$$\left(EI\tilde{\psi}' + J_c\ddot{\tilde{\psi}}\right)\Bigg|_{x=l} = -J_c v,$$

where

$$v = \left(J_0 + \int_0^l [(w_0)^2\rho - (w_0')^2\rho_2]\,dx + mw_0^2(l)\right)^{-1}$$

$$\times \Bigg\{ \tilde{M} + g\left(\int_0^l [(\rho w_0 - \rho_1 w_{0'})\cos\varphi_0 - \frac{1}{2}\rho_1(w_0')^2\sin\varphi_0]dx + mw_0(l)\cos\varphi_0\right)\tilde{\varphi}$$

$$+ g\int_0^l [(w_0'\cos\varphi_0 - \sin\varphi_0)\rho_1\tilde{w}' + \rho\tilde{w}\sin\varphi_0]dx + mg\tilde{w}(l,t) - EI\tilde{\psi}'(0,t) \Bigg\} + \cdots$$

(5.12)

Here, and in the sequel, the dots denote the terms of order 2 and higher with respect to $\tilde{\varphi}$, $\tilde{w}$, $\tilde{\psi}$ and their derivatives.

Since there is a one-to-one correspondence between the values of M and v for each

$$(\tilde{\varphi}(t), \dot{\tilde{\varphi}}(t), \tilde{w}(\cdot, t), \dot{\tilde{w}}(\cdot, t), \tilde{\psi}(\cdot, t), \dot{\tilde{\psi}}(\cdot, t)),$$

we will treat $v \in \mathbb{R}$ as a control parameter in (5.11).

Separation of variables. To construct finite dimensional approximations for the dynamical equations, we consider particular solutions of the linearized boundary value problem (5.11) as follows

$$\tilde{\varphi} = 0, \ v = 0, \ \tilde{w}(x, t) = \bar{w}(x)q(t), \ \tilde{\psi}(x, t) = \bar{\psi}(x)q(t).$$

Substitution of these relations into the linearized system (5.11) leads to the differential equation $\ddot{q}(t) = -\lambda q(t)$ together with the following spectral problem:

$$\bigl(K(\bar{\psi} - \bar{w}') + g\rho_1 \bar{w}' \sin \varphi_0\bigr)' - \lambda \rho \bar{w} = 0,$$

$$K(\bar{\psi} - \bar{w}') - (EI\bar{\psi}')' - \lambda I_\rho \bar{\psi} = 0, \ x \in (0, l),$$

$$\bar{w}(0) = 0, \ \bar{\psi}(0) = 0,$$

$$K(\bar{w}'(l) - \bar{\psi}(l)) - mg\bar{w}'(l) \sin \varphi_0 - m\lambda \bar{w}(l) = 0,$$

$$EI\bar{\psi}'(l) - \lambda J_c \bar{\psi}(l) = 0, \tag{5.13}$$

where λ is a scalar parameter.

To obtain the equations of motion with modal coordinates, let us first study the problem (5.13).

5.3 Spectrum of the Linearized Model

This section is devoted to the study of eigenvalues and eigenfunctions for the spectral problem (5.13)

Let us define

$$\mathcal{H} = \left\{ \begin{pmatrix} \bar{w} \\ \bar{\psi} \end{pmatrix} : \bar{w} \in H^1[0, l], \ \bar{\psi} \in H^1[0, l], \ \bar{w}(0) = \bar{\psi}(0) = 0 \right\},$$

where $H^1[0, l]$ is the Sobolev space. Consider the following symmetric bilinear form on $\mathcal{H}$:

$$\left\langle \begin{matrix} w_1 & w_2 \\ \psi_1 & \psi_2 \end{matrix} \right\rangle_{\mathscr{H}} = \int_0^l (\rho w_1 w_2 + I_\rho \psi_1 \psi_2)\, dx + m w_1(l) w_2(l) + J_c \psi_1(l) \psi_2(l). \quad (5.14)$$

The above definition implies the following result.

Lemma 5.1 *Let* (λ_1, w_1, ψ_1) *and* (λ_2, w_2, ψ_2) *be non-trivial solutions of the problem* (5.13). *If* $\lambda_1 \neq \lambda_2$ *then*

$$\left\langle \begin{matrix} w_1 & w_2 \\ \psi_1 & \psi_2 \end{matrix} \right\rangle_{\mathscr{H}} = 0.$$

Moreover, if $K(x) = \text{const}$ *and*

$$\frac{2\left(m + \int_0^l \rho\, dx\right) g \sin \varphi_0}{K} \leq 1, \quad \frac{K l^2}{E I} \leq 2 \quad (5.15)$$

then all eigenvalues of problem (5.13) *are non-negative real numbers.*

Proof Let (λ_1, w_1, ψ_1) be a solution of problem (5.13). Then

$$\lambda_1 \left\langle \begin{matrix} w_1 & w_2 \\ \psi_1 & \psi_2 \end{matrix} \right\rangle_{\mathscr{H}} = \left\langle \begin{matrix} \lambda_1 w_1 & w_2 \\ \lambda_1 \psi_1 & \psi_2 \end{matrix} \right\rangle_{\mathscr{H}}$$

$$= \int_0^l \left(K(\psi_1 - w_1') + g\rho_1 w_1' \sin \varphi_0 \right)' w_2\, dx$$

$$+ \int_0^l \left(K(\psi_1 - w_1') - (E I \psi_1')' \right) \psi_2\, dx$$

$$+ \left(K(w_1'(l) - \psi_1(l)) - m g w_1'(l) \sin \varphi_0 \right) w_2(l) + E I \psi_1'(l) \psi_2(l).$$

The integration by parts in the above expression gives

$$\lambda_1 \left\langle \begin{matrix} w_1 & w_2 \\ \psi_1 & \psi_2 \end{matrix} \right\rangle_{\mathscr{H}} = \int_0^l \{ K(w_1' w_2' + \psi_1 \psi_2 - w_1' \psi_2 - w_2' \psi_1)$$

$$+ E I \psi_1' \psi_2' - g\rho_1 w_1' w_2' \sin \varphi_0 \}\, dx. \quad (5.16)$$

By the permutation of arguments in (5.16), we get

$$\lambda_2 \left\langle \begin{matrix} w_1 & w_2 \\ \psi_1 & \psi_2 \end{matrix} \right\rangle_{\mathscr{H}} = \lambda_2 \left\langle \begin{matrix} w_2 & w_1 \\ \psi_2 & \psi_1 \end{matrix} \right\rangle_{\mathscr{H}} = \lambda_1 \left\langle \begin{matrix} w_1 & w_2 \\ \psi_1 & \psi_2 \end{matrix} \right\rangle_{\mathscr{H}}.$$

Thus, $\left\langle \begin{matrix} w_1 & w_2 \\ \psi_1 & \psi_2 \end{matrix} \right\rangle_{\mathcal{H}} = 0$ if $\lambda_1 \neq \lambda_2$. For $w_2 = w_1$ and $\psi_2 = \psi_1$, equality (5.16) implies

$$\lambda_1 \left\langle \begin{matrix} w_1 & w_1 \\ \psi_1 & \psi_1 \end{matrix} \right\rangle_{\mathcal{H}} = \int_0^l \left(K(w_1' - \psi_1)^2 + EI\psi_1'^2 - g\rho_1 w_1'^2 \sin\varphi_0 \right) dx$$

$$= \int_0^l \left(\frac{K}{2} w_1'^2 + EI\psi_1'^2 - K\psi_1^2 - g\rho_1 w_1'^2 \sin\varphi_0 \right) dx$$

$$+ \frac{1}{2} \int_0^l K(w_1' - 2\psi_1)^2 \, dx$$

$$\geq \int_0^l \left(\left(\frac{K}{2} - g\rho_1 \sin\varphi_0 \right) w_1'^2 + EI\psi_1'^2 - K\psi_1^2 \right) dx. \quad (5.17)$$

If the function $\psi_1(x)$ satisfies the boundary condition $\psi_1(0) = 0$ then the following Friedrichs–Wirtinger type inequality holds (see [4, 22]):

$$\int_0^l \psi_1^2(x)dx \leq \frac{l^2}{2} \int_0^l \psi_1'^2(x)dx.$$

Applying this inequality to (5.17), we conclude that

$$\lambda_1 \left\langle \begin{matrix} w_1 & w_1 \\ \psi_1 & \psi_1 \end{matrix} \right\rangle_{\mathcal{H}} \geq \int_0^l \left(\left(\frac{K}{2} - g\rho_1 \sin\varphi_0 \right) w_1'^2 + \left(EI - \frac{Kl^2}{2} \right) \psi_1'^2 \right) dx \geq 0$$

provided that conditions (5.15) hold. Therefore, all eigenvalues λ are non-negative real numbers.

In the rest of this section, we assume that K, EI, I_ρ, ρ are constants, and $\sin\varphi_0 = 0$. Under such assumptions, the coefficients of the spectral problem are constant, therefore, the general solution of the corresponding system of ordinary differential equations can be written explicitly. Such solution will be used for computing the coefficients of an approximate system in the next section.

Let us introduce the following dimensionless functions for the problem (5.13):

$$\zeta(x/l) = \bar{w}(x)/l, \quad \theta(x/l) = \bar{\psi}(x),$$

and parameters

$$p_1 = \frac{\rho l^2}{K}, \quad p_2 = \frac{K l^2}{EI}, \quad p_3 = \frac{I_\rho l^2}{EI}, \quad p_4 = \frac{ml}{K}, \quad p_5 = \frac{l J_c}{EI}.$$

Then (5.13) is reduced to the following problem:

$$\frac{d}{d\tau} \begin{pmatrix} \zeta(\tau) \\ \zeta_\tau(\tau) \\ \theta(\tau) \\ \theta_\tau(\tau) \end{pmatrix} = \begin{pmatrix} 0 & 1 & 0 & 0 \\ -\lambda p_1 & 0 & 0 & 1 \\ 0 & 0 & 0 & 1 \\ 0 & -p_2 & p_2 - \lambda p_3 & 0 \end{pmatrix} \times \begin{pmatrix} \zeta(\tau) \\ \zeta_\tau(\tau) \\ \theta(\tau) \\ \theta_\tau(\tau) \end{pmatrix}, \quad \tau = \frac{x}{l} \in (0, 1);$$

(5.18)

$$\zeta_\tau(1) - \theta(1) = \lambda p_4 \zeta(1), \quad \theta_\tau(1) = \lambda p_5 \theta(1), \quad \zeta(0) = \theta(0) = 0, \quad (5.19)$$

where $\zeta_\tau(\tau)$ and $\theta_\tau(\tau)$ stand for the derivatives with respect to τ.

To find solutions of the spectral problem, we compute the eigenvalues μ_j and eigenvectors v_j for the matrix in the right-hand side of (5.18):

$$\mu_j = i\sigma_j, \quad v_j = \begin{pmatrix} 4\mu_j(\lambda p_3 - \sigma_j^2) \\ \lambda c_3 + 4\lambda p_1(\sigma_j^2 - \lambda p_3) \\ \lambda c_3 \\ \lambda c_3 \mu_j \end{pmatrix}, \quad j = 1, 2, 3, 4,$$

where

$$\sigma_1 = -\sigma_2 = \frac{\sqrt{2}}{2} \sqrt{c_1 \lambda - \sqrt{c_2^2 \lambda^2 + c_3 \lambda}},$$

$$\sigma_3 = -\sigma_4 = \frac{\sqrt{2}}{2} \sqrt{c_1 \lambda + \sqrt{c_2^2 \lambda^2 + c_3 \lambda}}, \quad (5.20)$$

$$c_1 = p_1 + p_3, \quad c_2 = p_1 - p_3, \quad c_3 = 4 p_1 p_2.$$

Then the general solution of differential equations (5.18) is represented as

$$(\zeta, \zeta_\tau, \theta, \theta_\tau)^T (\tau) = C_1 v_1 e^{i\sigma_1 \tau} + C_2 v_2 e^{-i\sigma_1 \tau} + C_3 v_3 e^{i\sigma_3 \tau} + C_4 v_4 e^{-i\sigma_3 \tau}. \quad (5.21)$$

Note that a particular case $p_1 = p_2 = p_3 = 1$, $p_4 = p_5 = 0$ in considered in the paper [4]. The case of a Timoshenko beam with double eigenvalues is studied in the paper [23].

Substitution of (5.21) into the boundary conditions (5.19) leads to the system of linear algebraic equations with respect to C_1, C_2, C_3, C_4. That system has a non-trivial solution only if

$$\Delta(\lambda) = \begin{vmatrix} e^{-i\sigma_1} & e^{i\sigma_1} & e^{-i\sigma_3} & e^{i\sigma_3} \\ \sigma_1(\sigma_1^2 - \lambda p_3)e^{-i\sigma_1} & -\sigma_1(\sigma_1^2 - \lambda p_3)e^{i\sigma_1} & \sigma_3(\sigma_3^2 - \lambda p_3)e^{-i\sigma_3} & -\sigma_3(\sigma_3^2 - \lambda p_3)e^{i\sigma_3} \\ (\sigma_1^2 - \lambda p_3)(p_1 + ip_4\sigma_1) & (\sigma_1^2 - \lambda p_3)(p_1 - ip_4\sigma_1) & (\sigma_3^2 - \lambda p_3)(p_1 + ip_4\sigma_3) & (\sigma_3^2 - \lambda p_3)(p_1 - ip_4\sigma_3) \\ i\sigma_1 - \lambda p_5 & i\sigma_1 + \lambda p_5 & i\sigma_3 - \lambda p_5 & i\sigma_3 + \lambda p_5 \end{vmatrix} = 0.$$

$$(5.22)$$

The roots of the equation $\Delta(\lambda) = 0$ correspond to the eigenvalues λ of the spectral problem (5.13) with constant coefficients. As expression (5.23) define the analytic function $\Delta(\lambda)$, then the uniqueness theorem for analytic functions implies two possibilities: either $\Delta(\lambda) \equiv 0$ or the set of eigenvalues for the problem (5.13) is discrete.

For the case $p_4 = 0$, the computation of the determinant in formula (5.22) leads to the following characteristic equation:

$$2c_2^4 p_5 \{\sigma_1 \cos(\sigma_1) \sin(\sigma_3) + \sigma_3 \sin(\sigma_1) \cos(\sigma_3)\} \lambda^3$$

$$+ c_2^2 \Big\{ \sigma_1^2 \sin(\sigma_1) \sin(\sigma_3) c_2^2$$

$$+ 2\sigma_1 c_2 p_5 \cos(\sigma_1) \sin(\sigma_3) - 2c_2 \sigma_3 p_5 \sin(\sigma_1) \cos(\sigma_3)$$

$$+ \sigma_3^2 \sin(\sigma_1) \sin(\sigma_3) c_2^2 - 2\sigma_1 \sigma_3 c_2^2 + 4\sigma_1 p_5 \cos(\sigma_1) \sin(\sigma_3) c_3$$

$$+ 4\sigma_3 p_5 \sin(\sigma_1) \cos(\sigma_3) c_3 - 2\sigma_1 \sigma_3 \cos(\sigma_1) \cos(\sigma_3) c_2^2 \Big\} \lambda^2$$

$$+ 2 \Big\{ \sigma_1^2 \sin(\sigma_1) \sin(\sigma_3) c_2^2 c_3 - c_2 \sigma_3 p_5 \sin(\sigma_1) \cos(\sigma_3) c_3$$

$$+ \sigma_1 p_5 \cos(\sigma_1) \sin(\sigma_3) c_3^2 - 2\sigma_1 \sigma_3 \cos(\sigma_1) \cos(\sigma_3) c_2^2 c_3$$

$$+ \sigma_3 p_5 \sin(\sigma_1) \cos(\sigma_3) c_3^2 + \sigma_1 c_2 p_5 \cos(\sigma_1) \sin(\sigma_3) c_3$$

$$+ \sigma_3^2 \sin(\sigma_1) \sin(\sigma_3) c_2^2 c_3 - 2\sigma_1 \sigma_3 c_2^2 c_3 \Big\} \lambda$$

$$- 2\sigma_1 \sigma_3 \cos(\sigma_1) \cos(\sigma_3) c_3^2$$

$$+ \sigma_1^2 \sin(\sigma_1) \sin(\sigma_3) c_3^2 - \sigma_3^2 c_2^2 \sin(\sigma_1) \sin(\sigma_3) - 2\sigma_1 \sigma_3 c_2^2 \cos(\sigma_1) \cos(\sigma_3)$$

$$+ \sigma_3^2 \sin(\sigma_1) \sin(\sigma_3) c_3^2 - \sigma_1^2 c_2^2 \sin(\sigma_1) \sin(\sigma_3) - 2\sigma_1 \sigma_3 c_3^2 + 2\sigma_1 \sigma_3 c_2^2 = 0.$$

$$(5.23)$$

Let us consider a particular case of the motion without a payload: $p_4 = p_5 = 0$ (i.e. $m = 0$ and $J_c = 0$). Then characteristic equation (5.23) is reduced to

$$\left(1 - \kappa + \frac{c_3}{\lambda c_2^2}\right) \cos(\sigma_1 + \sigma_3) + \left(1 - \kappa - \frac{c_3 \kappa}{\lambda c_2^2}\right) \cos(\sigma_3 - \sigma_1) = \frac{c_3(\kappa - 1)}{\lambda c_2^2},$$

$$(5.24)$$

where

$$\kappa = \left(\frac{\sigma_1 - \sigma_3}{\sigma_1 + \sigma_3}\right)^2.$$

If $c_3 = 0$ then (5.22) is equivalent to the equation

$$\cos\left(\sqrt{p_1\lambda}\right) \cdot \cos\left(\sqrt{p_3\lambda}\right) = 0. \tag{5.25}$$

Below we study properties of the eigenvalues λ of the problem (5.18) and (5.19) under the assumption that all coefficients p_1, p_2, p_3, p_4, p_5 are positive constants. Taking into account that the beam with constant density satisfies the relation $I_\rho = \rho I/A$, we get

$$\frac{p_3}{p_1} = \frac{\gamma G}{E}.$$

In practice, the inequality $p_3 < p_1$ holds as $G/E \approx 0,37$ for metals used in manipulators, and $\gamma < 1$ (see, e.g., the paper [24] for formulas of the coefficient γ).

Distribution of the eigenvalues. Consider the Hilbert space

$$X = \{(\zeta, \theta, y, z) : \zeta, \theta \in L^2[0, 1], \ y, z \in \mathbb{C}\}$$

over the field of complex numbers. We define the inner products of elements $\xi_1 = (\zeta_1, \theta_1, y_1, z_1) \in X$ and $\xi_2 = (\zeta_2, \theta_2, y_2, z_2) \in X$ as

$$\langle \xi_1, \xi_2 \rangle = \int_0^1 \left(p_1 p_2 \zeta_1(\tau) \bar{\zeta}_2(\tau) + p_3 \theta_1(\tau) \bar{\theta}_2(\tau)\right) d\tau + p_2 p_4 y_1 \bar{y}_2 + p_5 z_1 \bar{z}_2. \tag{5.26}$$

Let A be a linear operator with the domain $D(A) \subset X$ and values in X:

$$D(A) = \{(\zeta, \theta, y, z) : \zeta, \theta \in H^2(0, 1), \ \zeta(0) = \theta(0) = 0, \ y = \zeta(1), \ z = \theta(1)\},$$

$$A(\zeta, \theta, y, z) = \left(\frac{\theta' - \zeta''}{p_1}, \frac{p_2(\theta - \zeta') - \theta''}{p_3}, \frac{\zeta'(1) - \theta(1)}{p_4}, \frac{\theta'(1)}{p_5}\right),$$

where $H^2(0, 1)$ is the Sobolev space.

It is easy to see that λ is an eigenvalue of the spectral problem (5.18) and (5.19) only if there is an element $\xi \in X$ satisfying the conditions

$$A\xi = \lambda\xi, \quad \xi \neq 0, \quad \xi \in D(A). \tag{5.27}$$

If λ and $\xi = (\zeta, \theta, y, z)$ satisfy (5.27) then functions $\zeta(\tau)$ and $\theta(\tau), \tau \in [0, 1]$ define the eigenmode of the beam corresponding to λ.

We prove now an important property of the operator A. Let $\xi_1, \xi_2 \in D(A)$. By performing the integration by parts with regard to the boundary conditions from $D(A)$, we get

$$\langle A\xi_1, \xi_2\rangle = \int_0^1 \left(p_2(\theta_1' - \zeta_1'')\bar\zeta_2 + (p_2(\theta_1 - \zeta_1') - \theta_1'')\bar\theta_2\right) d\tau$$

$$+ p_2(\zeta_1'(1) - \theta_1(1))\bar\zeta_2(1) + \theta_1'(1)\bar\theta_2(1)$$

$$= \int_0^1 \left(p_2\theta_1\bar\theta_2 + p_2(\zeta_1' - \theta_1)\bar\zeta_2' + (p_2\zeta_1 + \theta_1')\bar\theta_2'\right) d\tau$$

$$+ \left(p_2(\theta_1 - \zeta_1')\bar\zeta_2 - (p_2\zeta_1 + \theta_1')\bar\theta_2\right)\Big|_{\tau=0}^{1}$$

$$+ p_2(\zeta_1'(1) - \theta_1(1))\bar\zeta_2(1) + \theta_1'(1)\bar\theta_2(1)$$

$$= \int_0^1 \left(p_2(\theta_1\bar\theta_2 - \theta_1\bar\zeta_2' + \zeta_1\bar\theta_2' - \zeta_1\bar\zeta_2'') - \theta_1\bar\theta_2''\right) d\tau$$

$$+ \left(p_2\zeta_1\bar\zeta_2' + \theta_1\bar\theta_2'\right)\Big|_{\tau=0}^{1} - p_2\zeta_1(1)\bar\theta_2(1)$$

$$= \int_0^1 \left(p_2\zeta_1(\bar\theta_2' - \bar\zeta_2'') + \theta_1(p_2(\bar\theta_2 - \bar\zeta_2') - \bar\theta_2'')\right) d\tau$$

$$+ p_2\zeta_1(1)\left(\bar\zeta_2'(1) - \bar\theta_2(1)\right) + \theta_1(1)\bar\theta_2'(1)$$

$$= \langle \xi_1, A\xi_2\rangle.$$

Thus, A possesses properties of a self-adjoint operator [25]:

(1) *all eigenvalues λ are real;*
(2) *the eigenvectors $\xi_1, \xi_2 \in D(A)$, corresponding to distinct eigenvalues λ_1, λ_2, are orthogonal in X.*

The eigenfunctions $\zeta(\tau)$ and $\theta(\tau)$ of problem (5.18) and (5.19) can be chosen to be real as each eigenvalue λ is real. Hence, we will consider the elements of X with real components only and omit the sign of complex conjugation in the inner product (5.26).

Let us prove a lemma about the location of the eigenvalues.

Lemma 5.2 *If $p_2 < 2$ then all eigenvalues λ of the problem (5.18) and (5.19) belong to the semi-interval $[\lambda_0, +\infty)$, where*

$$\lambda_0 = \min\left\{\frac{1}{p_1 + 2p_4}, \frac{2 - p_2}{p_3 + 2p_5}\right\} > 0. \tag{5.28}$$

Proof Let $\xi \in D(A)$. Integrating by parts, we compute

$$\langle A\xi, \xi \rangle = \int_0^1 (p_2\zeta(\theta' - \zeta'') + p_2\theta(\theta - \zeta') - \theta\theta'')d\tau + p_2(\zeta'(1) - \theta(1))\zeta(1)$$

$$+ \theta(1)\theta'(1) = \int_0^1 (p_2(\zeta'^2 + \theta^2) + \theta'^2 - 2p_2\theta\zeta')d\tau$$

$$= \int_0^1 (p_2(\zeta' - \theta)^2 + \theta'^2)d\tau$$

$$= \int_0^1 \left(\frac{p_2}{2}\zeta'^2 + \theta'^2 - p_2\theta^2\right) d\tau + \frac{p_2}{2}\int_0^1 (\zeta' - 2\theta)^2 d\tau$$

$$\geq \int_0^1 \left(\frac{p_2}{2}\zeta'^2 + \theta'^2 - p_2\theta^2\right) d\tau. \tag{5.29}$$

As $\zeta(0) = 0$ and $\theta(0) = 0$, then functions $\zeta(\tau)$ and $\theta(\tau)$ satisfy the Friedrichs–Wirtinger type inequality [22]:

$$\int_0^1 \zeta'^2(\tau)d\tau \geq 2\int_0^1 \zeta^2(\tau)d\tau, \quad \int_0^1 \theta'^2(\tau)d\tau \geq 2\int_0^1 \theta^2(\tau)d\tau. \tag{5.30}$$

Besides, due to the Cauchy–Schwartz inequality,

$$\zeta^2(1) = \left(\int_0^1 1 \cdot \zeta'(\tau)d\tau\right)^2 \leq \int_0^1 d\tau \cdot \int_0^1 \zeta'^2(\tau)d\tau = \int_0^1 \zeta'^2(\tau)d\tau, \quad \theta^2(1)$$

$$\leq \int_0^1 \theta'^2(\tau)d\tau. \tag{5.31}$$

Let λ_0 be a positive number. Taking into account inequalities (5.29) and (5.31), we conclude that

$$\langle A\xi, \xi \rangle - \lambda_0\langle \xi, \xi \rangle \geq \int_0^1 \left(\frac{p_2}{2}\zeta'^2 + \theta'^2 - p_2\theta^2 - \lambda_0(p_1 p_2 \zeta^2 + p_3\theta^2)\right) d\tau$$

$$- \lambda_0(p_2 p_4 \zeta^2(1) + p_5\theta^2(1))$$

$$\geq \int_0^1 \left(p_2 \left(\frac{1}{2} - \lambda_0 p_4 \right) \zeta'^2 + (1 - \lambda_0 p_5) \theta'^2 \right.$$

$$\left. -(p_2 + \lambda_0 p_3) \theta^2 - \lambda_0 p_1 p_2 \zeta^2 \right) d\tau. \tag{5.32}$$

Now we estimate the expression (5.32) assuming that

$$\frac{1}{2} - \lambda_0 p_4 \geq 0, \ 1 - \lambda_0 p_5 \geq 0,$$

and using inequalities (5.30). We have

$$\langle A\xi, \xi \rangle - \lambda_0 \langle \xi, \xi \rangle \geq \int_0^1 \left(p_2(1 - \lambda_0(p_1 + 2p_4))\zeta^2(\tau) \right.$$

$$\left. +(2 - p_2 - \lambda_0(p_3 + 2p_5))\theta^2(\tau) \right) d\tau.$$

Hence,

$$\langle A\xi, \xi \rangle \geq \lambda_0 \langle \xi, \xi \rangle \tag{5.33}$$

for all $\xi \in D(A)$ provided that λ_0 is defined by expression (5.28) and $p_2 < 2$.

It is easy to show now that each eigenvalue λ of the operator A satisfies the inequality $\lambda \geq \lambda_0$. Indeed, if $A\xi = \lambda\xi$ for some $\xi \in D(A)$, $\xi \neq 0$, then $\langle A\xi, \xi \rangle = \lambda \langle \xi, \xi \rangle$. Using (5.33), we obtain

$$(\lambda - \lambda_0)\langle \xi, \xi \rangle \geq 0,$$

therefore, $\lambda - \lambda_0 \geq 0$ as $\xi \neq 0$.

As expression (5.22) defines the analytic function $\Delta(\lambda)$ which is not identically zero, the set of all eigenvalues of problem (5.18) and (5.19) is at most countable, does not have a finite accumulation point, and the multiplicity of each eigenvalue is finite. Moreover, we have the following result.

Lemma 5.3 *If $p_1 \neq p_3$ then the set of the roots of characteristic equation (5.22) is not bounded from the above.*

Proof By computing the determinant in (5.22), we obtain an asymptotic representation for $\Delta(\lambda)$ as follows:

$$\frac{\Delta(\lambda)}{\lambda^4} = -4p_1 p_4 p_5 (p_1 - p_3)^2 (\cos \sigma_1 \sin \sigma_3 + R(\lambda)), \tag{5.34}$$

where

$$R(\lambda) = \left(\frac{\sqrt{p_1}}{p_4} \cos \sigma_1 \cos \sigma_3 - \frac{\sqrt{p_3}}{p_5} \sin \sigma_1 \sin \sigma_3 \right) \frac{1}{\sqrt{\lambda}} + O\left(\frac{1}{\lambda}\right) \quad \text{as } \lambda \to +\infty.$$

Let us show that the function $\phi(\lambda) = \cos \sigma_1 \sin \sigma_3$ has a countable set of zeros on any semi-interval $\lambda \geq M$ (the values $\sigma_1 = \sigma_1(\lambda)$ and $\sigma_3 = \sigma_3(\lambda)$ depend on λ due to formulas (5.20)). It is easy to check that all solutions of the equation $\phi(\lambda) = 0$ can be represented as $\lambda = \lambda_k^*$ (k is an integer index), where λ_k^* is defined from the following conditions:

$$\sigma_1(\lambda_k^*) = \frac{\pi k}{2} \quad \text{if } k \text{ is odd,} \tag{5.35}$$

$$\sigma_3(\lambda_k^*) = \frac{\pi k}{2} \quad \text{if } k \text{ is even.} \tag{5.36}$$

Hence, λ_k^* can be computed explicitly by solving Eq. (5.20) with respect to λ:

$$\lambda = \frac{p_1 p_2 + (p_1 + p_3)\sigma_1^2 + \sqrt{\left((p_1 - p_3)\sigma_1^2 + p_1 p_2\right)^2 + 4p_1 p_2 p_3 \sigma_1^2}}{2p_1 p_3}, \tag{5.37}$$

$$\lambda = \frac{p_1 p_2 + (p_1 + p_3)\sigma_3^2 - \sqrt{\left((p_1 - p_3)\sigma_3^2 + p_1 p_2\right)^2 + 4p_1 p_2 p_3 \sigma_3^2}}{2p_1 p_3}. \tag{5.38}$$

By expanding formulas (5.37) and (5.38) into the Maclaurin series with respect to $1/\sigma_1$ and $1/\sigma_3$, we obtain the following asymptotic representations:

$$\lambda = \frac{p_1 p_2}{(p_1 - p_3)p_3} + \frac{\sigma_1^2}{p_3} + O(\sigma_1^{-2}) \quad \text{as } \sigma_1 \to \infty,$$

$$\lambda = \frac{p_2}{p_3 - p_1} + \frac{\sigma_3^2}{p_1} + O(\sigma_3^{-2}) \quad \text{as } \sigma_3 \to \infty. \tag{5.39}$$

By substituting expressions (5.35) into these formulas, we get an asymptotic representation of roots of the equation $\phi(\lambda) = 0$ for large values of k:

$$\lambda_k^* = \begin{cases} \dfrac{p_1 p_2}{(p_1 - p_3)p_3} + \dfrac{\pi^2 k^2}{4p_3} + O(1/k^2) & \text{if } k \text{ is odd,} \\[3mm] \dfrac{p_2}{p_3 - p_1} + \dfrac{\pi^2 k^2}{4p_1} + O(1/k^2) & \text{if } k \text{ is even.} \end{cases} \tag{5.40}$$

Thus, $\lambda_k^* \to +\infty$ as $k \to \infty$. Let us show now that the solutions set of the perturbed equation $\phi(\lambda) + R(\lambda) = 0$ is also not bounded above. Indeed, this fact follows from the continuity of functions $\phi(\lambda)$, $R(\lambda)$ and from the properties

$$\overline{\lim_{\lambda \to +\infty}} \; \phi(\lambda) > 0, \qquad \underline{\lim_{\lambda \to +\infty}} \; \phi(\lambda) < 0.$$

Hence, we deduce from formula (5.34) that there exist roots of the characteristic equation $\Delta(\lambda) = 0$ on any semi-interval of the form $\lambda \geq M$.

So, the set of all eigenvalues of the problem (5.18) and (5.19) is countable if $p_1 \neq p_3$. Let us arrange these values in the non-decreasing order by taking into account their multiplicities:

$$\lambda_1 \leq \lambda_2 \leq \cdots \leq \lambda_n \leq \cdots , \qquad \lambda_n \to +\infty \text{ as } n \to \infty.$$

It follows from Lemma 5.2 that $\lambda_1 \geq \lambda_0 > 0$ provided that $p_2 < 2$. Due to the representation (5.34), it is natural to assume that the value of λ_n is "close" to the corresponding value λ_k^* defined by formula (5.40) for large n.

To verify this assumption, we prove a result about the asymptotic distribution of the eigenfrequencies under additional assumptions on mechanical parameters.

Theorem 5.1 *Let $\sqrt{p_3/p_1} = r/q$, where $r < q$ are integers and q is odd. Then, for each eigenvalue λ_n of the problem (5.18) and (5.19), there are numbers $k = k(n)$ and*

$$\omega_k^* = \begin{cases} \dfrac{\pi k}{2p_3} & \textit{for odd } k, \\[2mm] \dfrac{\pi k}{2p_1} & \textit{for even } k, \end{cases} \tag{5.41}$$

such that $\sqrt{\lambda_n} = \omega_k^* + O(\frac{1}{k})$, $k \to \infty$ as $n \to \infty$.

Proof Let us express λ and σ_1 in σ_3 from the characteristic equation $\Delta(\lambda) = 0$ by formulas (5.20) and (5.39). Then by using (5.34), we get the following equation with respect to σ_3:

$$\cos \left(\sqrt{\frac{p_3}{p_1}} \sigma_3 - \frac{\sqrt{p_1}\, p_2(p_1 + p_3)}{2\sqrt{p_3}(p_1 - p_3)\sigma_3} \right) \left(\sin \sigma_3 + \frac{\sqrt{p_1 p_3}}{p_4 \sigma_3} \cos \sigma_3 \right)$$
$$= \frac{p_3}{p_5 \sigma_3} \sin \left(\sqrt{\frac{p_3}{p_1}} \sigma_3 - \frac{\sqrt{p_1}\, p_2(p_1 + p_3)}{2\sqrt{p_3}(p_1 - p_3)\sigma_3} \right) \sin \sigma_3 + O \left(\frac{1}{\sigma_3^2} \right). \tag{5.42}$$

Neglecting the higher order terms, we rewrite Eq. (5.42) in the form

$$\cos (\alpha \sigma) \sin \sigma = \varepsilon(\sigma), \qquad (\alpha = \sqrt{p_3/p_1}, \; \sigma = \sigma_3), \tag{5.43}$$

where $\varepsilon(\sigma) = O(1/\sigma)$ as $\sigma \to \infty$. It means that $\varepsilon(\sigma)$ admits the estimate

$$|\varepsilon(\sigma)| \leq \frac{M_1}{\sigma} \quad \text{as } \sigma \geq \tilde{\sigma} \tag{5.44}$$

with some positive constants $\tilde{\sigma}$ and M_1. The function

$$f(\sigma) = \sin \sigma \cos(\alpha \sigma)$$

vanishes at $\sigma = \sigma_k^*$, where

$$
\sigma_k^* =
\begin{cases}
\dfrac{\pi k}{2} & \text{if } k \text{ is even,} \\[2mm]
\dfrac{\pi k}{2\alpha} & \text{if } k \text{ is odd.}
\end{cases}
\tag{5.45}
$$

To localize the solutions of Eq. (5.43), we apply Taylor's formula with the Lagrange form of the remainder:

$$
f(\sigma_k^* + d) = f'(\sigma_k^*)d + \frac{f''(\eta)}{2}d^2, \quad |\eta - \sigma_k^*| \le |d|,
\tag{5.46}
$$

$$
f'(\sigma_k^*) =
\begin{cases}
(-1)^{k/2} \cos\left(\dfrac{\pi k \alpha}{2}\right) & \text{if } k \text{ is even,} \\[3mm]
(-1)^{(k+1)/2} \alpha \sin\left(\dfrac{\pi k}{2\alpha}\right) & \text{if } k \text{ is odd,}
\end{cases}
\tag{5.47}
$$

$$f''(\eta) = -(1 + \alpha^2) \sin \eta \cos(\alpha \eta) - 2\alpha \cos \eta \sin(\alpha \eta).$$

The condition $\alpha = r/q$ (q is odd) implies the periodicity of $f(\sigma)$ and the property $f'(\sigma_k^*) \ne 0$ for all integers k. Let us prove this property by assuming the contrary. If $f'(\sigma_k^*) = 0$ for some k then one of the following alternatives holds because of (5.47):

(1) k is even, kr/q is odd integer;
(2) k is odd, kq/r is even integer.

Neither of these alternatives is possible for odd q. Thus, the values of $|f'(\sigma_k^*)|$ are separated from zero under the conditions of this Theorem[1]:

$$M_2 = \inf_{k \in \mathbb{Z}} |f'(\sigma_k^*)| > 0.$$

Let us denote

$$h = \frac{M_2}{2(\alpha + 1)^2} > 0$$

and check that the function $f(\sigma)$ is strictly monotone on each interval of the form $I_k = (\sigma_k^* - h, \sigma_k^* + h)$. Indeed, the derivative $f'(\sigma)$ is continuous and non-vanishing for $\sigma \in I_k$:

---

[1] It is easy to show that, if α is irrational or $\alpha = r/q$, $(r, q) = 1$, and q is even, then the lower limit of $|f'(\sigma_k^*)|$ is equal to zero.

$$|f'(\sigma)| = |f'(\sigma_k^*) + f''(\eta)(\sigma - \sigma_k^*)| \geq |f'(\sigma_k^*)| - h \sup_{\eta \in \mathbb{R}} |f''(\eta)|$$

$$\geq M_2 - h(1+\alpha)^2 = \frac{M_2}{2} > 0.$$

Since σ_k^* are located in intervals I_k and the continuous function $f(\sigma)$ is periodic, the values of $f(\sigma)$ are separated from zero on the set $S = \mathbb{R}\backslash\bigcup_{k\in\mathbb{Z}} I_k$:

$$M_3 = \inf_{\sigma \in S} |f(\sigma)| > 0.$$

Let

$$\sigma_{min} = \max\left\{\tilde{\sigma}, \frac{M_1}{M_3}, \sqrt{\frac{2M_1}{M_2}}\right\}. \tag{5.48}$$

For such a choice of σ_{min}, there are no solutions of Eq. (5.43) in the interval $\sigma \in (\sigma_{min}, +\infty) \cap S$. Indeed,

$$|f(\sigma)| \geq M_3 \geq \frac{M_1}{\sigma_{min}} > \frac{M_1}{\sigma} \geq |\varepsilon(\sigma)| \quad \text{if } \sigma \in S, \ \sigma > \sigma_{min}.$$

Therefore, each solution of Eq. (5.43) from the interval $\sigma > \sigma_{min}$ belongs to some I_k. Vice versa, for k large enough, each interval $I_k \subset (\sigma_{min}, +\infty)$ contains a solution of Eq. (5.43). To prove this fact, let us consider an arbitrary σ_k^* satisfying the inequality $\sigma_k^* > \sigma_{min} + d$. Let us construct a segment $[\sigma_k^* - d, \sigma_k^* + d]$ such that the function $f(\sigma) - \varepsilon(\sigma)$ is either identically zero or takes distinct signs at the ends of this segment. For this purpose it suffices to choose a number d satisfying the conditions

$$|f(\sigma_k^* - d)| \geq |\varepsilon(\sigma_k^* - d)|, \ |f(\sigma_k^* + d)| \geq |\varepsilon(\sigma_k^* + d)|, \ 0 < d < h, \tag{5.49}$$

because the function $f(\sigma)$ changes its sign at σ_k^*. Formula (5.46) implies that conditions (5.49) hold if

$$|f(\sigma_k^* \pm d)| \geq |f'(\sigma_k^*)|d - \frac{d^2}{2}\sup_{\eta\in\mathbb{R}}|f''(\eta)| \geq M_2 d - \frac{(\alpha+1)^2}{2}d^2 \geq \frac{M_1}{\sigma_k^* - d}. \tag{5.50}$$

The mean value theorem implies the following inequality:

$$\frac{1}{\sigma_k^* - d} < \frac{1}{\sigma_k^*} + \frac{d}{\sigma_{min}^2}, \quad (0 < d < \sigma_k^* - \sigma_{min}). \tag{5.51}$$

Hence, condition (5.50) is satisfied if

$$d^2 - 2bd + c \leq 0, \quad 0 < d < h, \tag{5.52}$$

where

$$b = \frac{M_2 - M_1/\sigma_{min}^2}{(\alpha + 1)^2}, \qquad c = \frac{2M_1}{(\alpha + 1)^2\sigma_k^*} > 0.$$

Let us note that $b > 0$ due to the inequality $\sigma_{min}^2 \geq 2M_1/M_2$ obtained from formula (5.48). By solving inequality (5.52), we conclude that the segment $[\sigma_k^* - d, \sigma_k^* - d]$ contains a solution of Eq. (5.43) provided that $d = d_k^*$:

$$d_k^* = b - \sqrt{b^2 - c} = \frac{M_1}{(M_2 - M_1/\sigma_{min}^2)\sigma_k^*} + O\left(\frac{1}{(\sigma_k^*)^2}\right), \quad \sigma_k^* > \frac{2M_1}{b^2(\alpha + 1)^2}. \tag{5.53}$$

It remains to show that the set $I_k \setminus [\sigma_k^* - d, \sigma_k^* - d]$ does not contain any solution of Eq. (5.43) for $d = d_k^*$. Taking into account the estimate (5.44), it suffices to prove that

$$|f(\sigma_k^* \pm \delta)| > \frac{M_1}{\sigma_k^* - \delta} \tag{5.54}$$

for any $\delta \in (d_k^*, h]$. Formulas (5.46) and (5.51) imply that inequality (5.54) holds if $P(\delta) = \delta^2 - 2b\delta + c < 0$. As the roots of the quadratic polynomial $P(\delta)$ are equal to $\delta = d_k^*$ and $\delta = b + \sqrt{b^2 - c} > b \geq h$, then $P(\delta)$ is of negative sign for each $\delta \in (d_k^*, h]$.

Thus, each sufficiently large σ satisfying Eq. (5.43) is contained in the closed neighborhood of some σ_k^* with radius d_k^*. Formula (5.53) implies the estimate $\sigma = \sigma_k^* + O(1/k)$ for the solutions of Eq. (5.43) as $k \to \infty$. By substituting this estimate into formula (5.38) and extracting the root, we obtain the following estimate:

$$\sqrt{\lambda_n} = \omega_k^* + O(1/k).$$

5.4 Galerkin's Approximations

In order to construct Galerkin's approximations, we use the following variational formulation of the dynamical equations (cf. [26]): if

$$\left(\tilde{\varphi}(t), \tilde{w}(x, t), \tilde{\psi}(x, t)\right), \quad 0 \leq x \leq l$$

is a solution of the boundary value problem (5.11) with control $M(t)$ on $t \in \mathscr{I} \subset \mathbb{R}$, then

$$\ddot{\tilde{\varphi}}(t) - v = 0,$$

$$\tilde{\mu} = \int_0^l \delta\tilde{w}(x,t)\{(\ddot{\tilde{w}} + xv - g\tilde{\varphi}\sin\varphi_0)\rho + \left(K(\tilde{\psi} - \tilde{w}') + \rho_1 g(\tilde{w}'\sin\varphi_0\right.$$

$$\left. + \tilde{\varphi}w_0'\cos\varphi_0)\right)' + \cdots\}dx + \int_0^l \delta\tilde{\psi}(x,t)\{I_\rho\ddot{\tilde{\psi}} + K(\tilde{\psi} - \tilde{w}') - (EI\tilde{\psi}')'$$

$$+ I_\rho v\}\,dx + \delta\tilde{\psi}(l,t)\left\{J_c\ddot{\tilde{\psi}} + EI\tilde{\psi}' + J_c v\right\}|_{x=l}$$

$$+ \delta\tilde{w}(l,t)\{K(\tilde{w}' - \tilde{\psi}) + m(\ddot{\tilde{w}} + lv - g(\tilde{\varphi} + \tilde{w}')\sin\varphi_0$$

$$- g\tilde{\varphi}w_0'\cos\varphi_0) + \cdots\}\Big|_{x=l} = 0, \quad \forall t \in \mathcal{I}, \tag{5.55}$$

for any admissible variations $(\delta\tilde{w}(x,t), \delta\tilde{\psi}(x,t))$ satisfying the boundary conditions $\delta\tilde{w}(0,t) = 0$ and $\delta\tilde{\psi}(0,t) = 0$. Expression (5.55) is obtained from (5.7) by using the Maclaurin series expansion and neglecting the higher order terms. The values v and M are related by formula (5.12).

Let us fix an integer number $N \geq 1$ and consider non-trivial solutions (λ_j, w_j, ψ_j) of the problem (5.13) for $j = 1, 2, \ldots, N$. Suppose that all λ_j are distinct and substitute the finite sums

$$\tilde{w}(x,t) = \sum_{j=1}^N q_j(t)w_j(x), \quad \tilde{\psi}(x,t) = \sum_{j=1}^N q_j(t)\psi_j(x) \tag{5.56}$$

into expression (5.55). Suppose also that the variational form (5.55) vanishes for all variations $\delta\tilde{w}$ and $\delta\tilde{\psi}$ from finite dimensional spaces

$$\delta\tilde{w}(\cdot, t) \in \text{span}\{w_1(\cdot), \ldots, w_N(\cdot)\}, \quad \delta\tilde{\psi}(\cdot, t) \in \text{span}\{\psi_1(\cdot), \ldots, \psi_N(\cdot)\}.$$

Assuming $\delta\tilde{w}(x,t) = w_i(x)$ and $\delta\tilde{\psi}(x,t) = \psi_i(x)$ in formula (5.55) for $i = 1, \ldots, N$ and applying Lemma 5.1, we get the following system of ordinary differential equations with respect to $\tilde{\varphi}, q_1, q_2, \ldots, q_N$:

$$\ddot{\tilde{\varphi}} = v, \quad \ddot{q}_j = -\lambda_j q_j + a_j\tilde{\varphi} - b_j v + \cdots, \quad j = 1, 2, \ldots, N, \tag{5.57}$$

where

$$a_j = g\frac{\left(\int_0^l \rho w_j\,dx + mw_j(l)\right)\sin\varphi_0 + \int_0^l \rho_1 w_0' w_j'\,dx\cos\varphi_0}{\int_0^l (\rho w_j^2 + I_\rho\psi_j^2)\,dx + mw_j^2(l) + J_c\psi_j^2(l)},$$

$$b_j = \frac{\int_0^l (\rho x w_j + I_\rho\psi_j)dx + mlw_j(l) + J_c\psi_j(l)}{\int_0^l (\rho w_j^2 + I_\rho\psi_j^2)\,dx + mw_j^2(l) + J_c\psi_j^2(l)}, \tag{5.58}$$

and dots denote the nonlinear terms with respect to the state variables in system (5.57).

When studying the observability and stabilization problems for the model considered, one should take into account relation (5.12) between the control parameter v and the perturbed control torque $\tilde{M} = M - M_0$. By substituting expression (5.56) into formula (5.12), we get an approximate relation between v and $\tilde{M}$ for a system with finite number of degrees of freedom:

$$v = u + d_0\tilde{\varphi} + \sum_{j=1}^{N} d_j q_j + \cdots, \tag{5.59}$$

where

$$u = \frac{\tilde{M}}{J_0 + \int_0^l (w_0^2\rho - w_0'^2 \rho_2)dx + mw_0^2(l)},$$

$$d_0 = \frac{\int_0^l [(\rho w_0 - \rho_1 w_0')\cos\varphi_0 - \frac{1}{2}(w_0')^2\rho_1 \sin\varphi_0]dx + mw_0(l)\cos\varphi_0}{J_0 + \int_0^l (w_0^2\rho - w_0'^2 \rho_2)dx + mw_0^2(l)}g,$$

$$d_j = \frac{g\int_0^l [(w_0'\cos\varphi_0 - \sin\varphi_0)\rho_1 w_j' + \rho w_j \sin\varphi_0]dx + mgw_j(l)\sin\varphi_0 + EI\psi_j'(0)}{J_0 + \int_0^l (w_0^2\rho - w_0'^2 \rho_2)dx + mw_0^2(l)}.$$

By applying relation (5.59), system (5.57) can be represented in the following matrix form:

$$\begin{aligned}\dot{z}_1 &= A_{11}z_1 + A_{12}z_2 + B_1 u + R_1(z, u),\\ \dot{z}_2 &= A_{21}z_1 + A_{22}z_2 + B_2 u + R_2(z, u), \quad z = (z_1^T, z_2^T)^T,\end{aligned} \tag{5.60}$$

where z is the state vector and u is the control,

$$z_1 = (\tilde{\varphi}, \dot{\tilde{\varphi}})^T, \quad z_2 = (q_1, \dot{q}_1, q_2, \dot{q}_2, \ldots, q_N, \dot{q}_N)^T,$$

$$u = \frac{\tilde{M}}{J_0 + \int_0^l (w_0^2\rho - w_0'^2 \rho_2)dx + mw_0^2(l)}.$$

As we can see, the control u is proportional to the deviation of the torque M from its reference value M_0 for system (5.60). Thus, to formulate problems of observation and stabilization with respect to the available measurements for a real manipulator, we will use system (5.60) instead of the simplified model (5.57). The matrices with coefficients of system (5.60) have the following forms:

$$A_{11} = \begin{pmatrix} 0 & 1 \\ d_0 & 0 \end{pmatrix}, \quad A_{12} = \begin{pmatrix} 0 & 0 & 0 & 0 & \dots & 0 & 0 \\ d_1 & 0 & d_2 & 0 & \dots & d_N & 0 \end{pmatrix},$$

$$B_1 = \begin{pmatrix} 0 \\ 1 \end{pmatrix}, \quad B_2 = (0, -b_1, \dots, 0, -b_N)^T,$$

$$A_{21} = \begin{pmatrix} 0 & 0 \\ a_1 - b_1 d_0 & 0 \\ 0 & 0 \\ a_2 - b_2 d_0 & 0 \\ \vdots & \vdots \\ 0 & 0 \\ a_N - b_N d_0 & 0 \end{pmatrix},$$

$$A_{22} = \begin{pmatrix} 0 & 1 & 0 & 0 & \dots & 0 & 0 \\ -\lambda_1 - b_1 d_1 & 0 & -b_1 d_2 & 0 & \dots & -b_1 d_N & 0 \\ 0 & 0 & 0 & 1 & \dots & 0 & 0 \\ -b_2 d_1 & 0 & -\lambda_2 - b_2 d_2 & 0 & \dots & -b_2 d_N & 0 \\ \vdots & \vdots & \vdots & \vdots & \ddots & \vdots & \vdots \\ 0 & 0 & 0 & 0 & \dots & 0 & 1 \\ -b_N d_1 & 0 & -b_N d_2 & 0 & \dots & -\lambda_N - b_N d_N & 0 \end{pmatrix},$$

and the nonlinear term $R(z, u) = (R_1^T, R_2^T)^T$ satisfies the estimate

$$\|R(z, u)\| = O(\|z\|^2 + u^2)$$

in a neighborhood of the equilibrium $z = 0$, $u = 0$.

System (5.60) is considered as a finite dimensional approximation of the boundary value problem (5.11) with flexible modes of up to the Nth order.

5.5 Controllability and Stabilization by a State Feedback Law

As the values of control parameters v and u in systems (5.57) and (5.60) are connected by the reversible transformation (5.59), the study of local controllability for system (5.60) is reduced to the analysis of system (5.57).

Let us write the linear approximation of system (5.57) around zero in the matrix form:

$$\dot{z} = \bar{A}z + \bar{B}v, \tag{5.61}$$

$$
z = \begin{pmatrix} \tilde{\varphi} \\ \dot{\tilde{\varphi}} \\ q_1 \\ \dot{q}_1 \\ \vdots \\ q_N \\ \dot{q}_N \end{pmatrix}, \quad
\bar{A} = \begin{pmatrix}
0 & 1 & 0 & 0 & \dots & 0 & 0 \\
0 & 0 & 0 & 0 & \dots & 0 & 0 \\
0 & 0 & 0 & 1 & \dots & 0 & 0 \\
a_1 & 0 & -\lambda_1 & 0 & \dots & 0 & 0 \\
\vdots & \vdots & \vdots & \vdots & \ddots & \vdots & \vdots \\
0 & 0 & 0 & 0 & \dots & 0 & 1 \\
a_N & 0 & 0 & 0 & \dots & -\lambda_N & 0
\end{pmatrix}, \quad
\bar{B} = \begin{pmatrix} 0 \\ 1 \\ 0 \\ -b_1 \\ \vdots \\ 0 \\ -b_N \end{pmatrix}.
$$

The following controllability criterion holds.

Lemma 5.4 *System (5.61) is controllable iff all eigenvalues λ_j are distinct and $a_j + \lambda_j b_j \neq 0$ for all $j = 1, 2, \dots, N$.*

Proof Straightforward computations show that

$$
\det(\bar{B}, \bar{A}\bar{B}, \dots, \bar{A}^{2N+1}\bar{B}) = (-1)^{N+1} \cdot \prod_{1 \leq j \leq N} (a_j + \lambda_j b_j)^2 \cdot \prod_{1 \leq i < k \leq N} (\lambda_i - \lambda_k)^2.
$$

Therefore, the linear system (5.61) is completely controllable by the Kalman criterion (see, e.g., [27, 28]). ∎

The above statement can be easily extended to the system with the control parameter u. We state here a local version of this result for nonlinear system (5.60).

Theorem 5.2 *System (5.60) is locally controllable in a neighborhood of the origin if all eigenvalues λ_j are distinct and $a_j + \lambda_j b_j \neq 0$ for all $j = 1, 2, \dots, N$.*

Proof System (5.61) is completely controllable by Lemma 5.4. As for each value of the state vector z there is a one-to-one correspondence between values v and u in formula (5.59), we conclude that the linear approximation of system (5.60) is also completely controllable. Therefore, system (5.60) is locally controllable in a neighborhood of the origin due to the theorem on controllability by the linear approximation (see, e.g., [27]). ∎

We propose an explicit control design scheme below in order to stabilize system (5.60) by a state feedback law.

Lemma 5.5 *Suppose that all eigenvalues $(\lambda_1, \dots, \lambda_N)$ are positive and distinct. Suppose also that $a_j + \lambda_j b_j \neq 0$ for all $j = \overline{1, N}$. Then there exist positive constants k_1, k_2, h_0 such that the feedback control*

$$
v(\tilde{\varphi}, \dot{\tilde{\varphi}}, q_1, \dots, q_N) = \frac{k_1 \tilde{\varphi} + h_0 \dot{\tilde{\varphi}} + \sum_{j=1}^{N} \left(a_j b_j \tilde{\varphi} - (a_j + \lambda_j b_j) q_j \right)}{\sum_{j=1}^{N} b_j^2 - k_2}, \quad (5.62)
$$

ensures the asymptotic stability of the trivial solution of system (5.61).

Proof Let us consider the following quadratic form with respect to the state variables of system (5.61):

$$2V(z) = k_1\tilde{\varphi}^2 + k_2\dot{\tilde{\varphi}}^2 + \sum_{j=1}^{N}\left(\lambda_j q_j{}^2 + \dot{q}_j^2 - 2a_j\tilde{\varphi}q_j + 2b_j\dot{\tilde{\varphi}}\dot{q}_j\right). \qquad (5.63)$$

It can be shown that the above form is positive definite for some values of constants k_1 and k_2. Indeed, by applying the Cauchy–Schwartz inequality we get

$$2V \geq G_1\left(-|\tilde{\varphi}|, \left(\sum_{j=1}^{N}\lambda_j q_j^2\right)^{1/2}\right) + G_2\left(-|\dot{\tilde{\varphi}}|, \left(\sum_{j=1}^{N}\dot{q}_j^2\right)^{1/2}\right), \qquad (5.64)$$

where

$$G_1(\alpha, \beta) = k_1\alpha^2 + 2\left(\sum_{j=1}^{N}\frac{a_j{}^2}{\lambda_j}\right)^{1/2}\alpha\beta + \beta^2,$$

$$G_2(\alpha, \beta) = k_2\alpha^2 + 2\left(\sum_{j=1}^{N}b_j{}^2\right)^{1/2}\alpha\beta + \beta^2.$$

According to the Sylvester criterion, both quadratic forms G_1 and G_2 are positive definite provided that

$$k_1 > \sum_{j=1}^{N}\frac{a_j{}^2}{\lambda_j}, \quad k_2 > \sum_{j=1}^{N}b_j{}^2. \qquad (5.65)$$

Suppose that k_1 and k_2 satisfy the above inequalities. Then estimate (5.64) implies that the form V is positive definite.

The time derivative of V along the trajectories of system (5.61) has the form

$$\dot{V} = \left(k_2 - \sum_{j=1}^{N}b_j{}^2\right)\dot{\tilde{\varphi}}v + \left(k_1\tilde{\varphi} + \sum_{j=1}^{N}(a_jb_j\tilde{\varphi} - (a_j + \lambda_jb_j)q_j)\right)\dot{\tilde{\varphi}}.$$

Let h_0 be an arbitrary positive constant. We define a feedback control from the condition

$$\dot{V} = -h_0\dot{\tilde{\varphi}}^2 \qquad (5.66)$$

along the trajectories of the closed-loop system. Thus we get expression (5.62) for the feedback control.

Let us apply the Barbashin–Krasovskii theorem [29] (or LaSalle's invariance principle) to prove the asymptotic stability of the trivial solution of the closed-loop system (5.61), (5.62). Consider the set

$$Z_0 = \{(\tilde{\varphi}, \dot{\tilde{\varphi}}, q_1, \ldots, \dot{q}_N) \in \mathbb{R}^{2N+2} \mid \dot{V} = 0\}.$$

Each positive semi-trajectory of system (5.61) and (5.62) on Z_0 satisfies the following relations:

$$\ddot{q}_j = -\lambda_j q_j + a_j \tilde{\varphi},$$

$$\sum_{j=1}^{N} [-a_j b_j \tilde{\varphi} + (a_j + \lambda_j b_j) q_j] = k_1 \tilde{\varphi} = \text{const}, \quad t \geq 0.$$

Hence,

$$\sum_{j=1}^{N} (a_j + \lambda_j b_j) \left(A_j \cos(\sqrt{\lambda_j} t) + B_j \sin(\sqrt{\lambda_j} t) \right) = \left(k_1 - \sum_{j=1}^{N} \frac{a_j^2}{\lambda_j} \right) \tilde{\varphi} \quad (5.67)$$

for some constants A_j, B_j, and $\tilde{\varphi}$. As the functions

$$\left\{ 1, \sin(\sqrt{\lambda_j} t), \cos(\sqrt{\lambda_j} t) \right\}_{j=1}^{N}$$

form a linearly independent system on $[0, +\infty)$ by Theorem 1.2.17 of [30], we conclude that (5.67) holds only if $\tilde{\varphi} = 0$ and $A_j = B_j = 0$ for all $j = 1, 2, \ldots, N$. Therefore, the set Z_0 contains only the trivial trajectory $x = 0$ of the closed-loop system (5.61) and (5.62). So the trivial solution is asymptotically stable by the Barbashin–Krasovskii theorem [29].

Let us now apply the above result to nonlinear system (5.60).

Theorem 5.3 *Suppose that all eigenvalues* $(\lambda_1, \ldots, \lambda_N)$ *are positive and distinct. Suppose also that* $a_j + \lambda_j b_j \neq 0$ *for all* $j = \overline{1, N}$. *Then there is a feedback control* $u = Kz$ *ensuring the asymptotic stability of the trivial solution of system* (5.60). *This control is defined by formulas*

$$u = Kz, \quad K = (K_1, K_2), \quad K_1 = \left(-d_0 - \frac{h_1 + \sum_{j=1}^{N} a_j (b_j + a_j/\lambda_j)}{h_2}, -\frac{h_0}{h_2} \right), \quad (5.68)$$

$$K_2 = \left(-d_1 + \frac{a_1 + \lambda_1 b_1}{h_2}, 0, -d_2 + \frac{a_2 + \lambda_2 b_2}{h_2}, 0, \ldots, -d_N + \frac{a_N + \lambda_N b_N}{h_2}, 0 \right),$$

where h_0, h_1, h_2 *are arbitrary positive constants.*

Proof For arbitrary positive constants h_1 and h_2, we put

$$k_1 = h_1 + \sum_{j=1}^{N} \frac{a_j^2}{\lambda_j}, \quad k_2 = h_2 + \sum_{j=1}^{N} b_j^2.$$

Then inequalities (5.65) hold, and the Lyapunov function $V(z)$ (5.63) is positive definite. Therefore, for any constant $h_0 > 0$, the control (5.62) ensures the asymptotic stability of the trivial solution of system (5.61) by Lemma 5.5. By using relation (5.59) in formula (5.62) and neglecting the nonlinear terms, we obtain the formula $u = Kz$ of type (5.68). Thus, the trivial solution of the system of linear approximation for (5.60) with $u = Kz$ is asymptotically stable by Lemma 5.5. This fact implies the asymptotic stability of the trivial solution of system (5.60) with $u = Kz$ by the theorem on stability by the linear approximation.

5.6 Observability Conditions

In practice, the operating modes of real manipulators do not allow to measure the values of functions $w(x, t)$ and $\psi(x, t)$ at each point $x \in [0, l]$. Therefore, the state vector of system (5.60) should be estimated by solving the observation problem with respect to available output signals. Such output signals may be generated in practice by a set of strain gauges located at some point of the manipulator with the lagrangian coordinate $x = l_0, 0 \le l_0 \le l$ (see Fig. 5.1). Thus we assume that a sensor measures the value of $\psi'(x, t)\big|_{x=l_0}$ for $x = l_0$ and all $t \ge 0$. In the papers [31, 32], the stabilization problem with an output feedback law has been considered for the Euler–Bernoulli beam with a free end in the case $l_0 = 0$ (collocated sensor and actuator). Note that we will consider here a general non-collocated case ($l_0 \ne 0$) for the Timoshenko beam model.

By computing the difference of functions $\varphi(t)$, $\psi'(x, t)\big|_{x=l_0}$ with their reference values corresponding to equilibrium (5.9), we define the output of system (5.60) as follows:

$$y_1(t) = \tilde{\varphi}(t), \quad y_2(t) = \sum_{j=1}^{N} \psi_j{}'(l_0)q_j(t). \tag{5.69}$$

Let us rewrite expressions (5.69) in the form

$$y_1 = C_1 z_1, \quad y_2 = C_2 z_2, \quad C_1 = (1, 0), \quad C_2 = (\chi_1, 0, \chi_2, 0, \dots, \chi_N, 0), \tag{5.70}$$

where $\chi_j = \psi_j{}'(l_0)$.

The purpose of this section is to solve the observability problem, i.e. to estimate the state vector $z(t)$ of system (5.60) for available information about the values of u and y. Note that the observability of the Euler–Bernoulli beam has been studied in the monograph [33, Chap. 5].

We will use the following sufficient conditions for the observability.

Theorem 5.4 *System (5.60) is locally observable at $z = 0$ with respect to the output (5.70) provided that*

$$
\begin{vmatrix}
\pi_{11} & \pi_{12} & \cdots & \pi_{1N} \\
\pi_{21} & \pi_{22} & \cdots & \pi_{2N} \\
\vdots & \vdots & \ddots & \vdots \\
\pi_{N1} & \pi_{N2} & \cdots & \pi_{NN}
\end{vmatrix}
\neq 0,
\tag{5.71}
$$

where $\pi_{1,j} = \chi_j$, $\pi_{k,j} = -\lambda_j \pi_{k-1,j} - d_j \sum_{i=1}^{N} \pi_{k-1,i} b_i$, $j = \overline{1, N}$, $k = \overline{2, N}$.
In particular, for $N = 1$, condition (5.71) is equivalent to the inequality $\chi_1 \neq 1$, and, for $N = 2$, to the following one:

$$
\chi_1 \chi_2 (\lambda_1 - \lambda_2 + b_1 d_1 - b_2 d_2) + b_2 \chi_2^2 d_1 - b_1 \chi_1^2 d_2 \neq 0.
$$

Proof By using the output y_1 we write the linear part of system (5.60) and (5.70) as

$$
\begin{aligned}
z_1 &= (y_1, \dot{y}_1)^T, \\
\dot{z}_2 &= A_{22} z_2 + B_2 u + (0, a_1 - b_1 d_0, \ldots, 0, a_N - b_N d_0)^T y_1, \\
y_2 &= C_2 z_2.
\end{aligned}
\tag{5.72}
$$

Hence, the components $z_1(t)$ is uniquely defined from the function y_1, and the subsystem with respect to z_2 is observable provided that the pair (A_{22}, C_2) satisfies the observability rank condition (see, e.g., [27, 28]):

$$
\text{rank}
\begin{pmatrix}
C_2 \\
C_2 A_{22} \\
\vdots \\
C_2 A_{22}^{2N-1}
\end{pmatrix}
= 2N.
$$

Straightforward computations show that

$$
\det
\begin{pmatrix}
C_2 \\
C_2 A_{22} \\
\vdots \\
C_2 A_{22}^{2N-1}
\end{pmatrix}
=
\begin{vmatrix}
\pi_{11} & \pi_{12} & \cdots & \pi_{1N} \\
\pi_{21} & \pi_{22} & \cdots & \pi_{2N} \\
\vdots & \vdots & \ddots & \vdots \\
\pi_{N1} & \pi_{N2} & \cdots & \pi_{NN}
\end{vmatrix}^2 .
$$

Therefore, relation (5.71) yields the observability rank condition for the linearized system (5.60) and (5.70). Hence, nonlinear system (5.60) and (5.70) is strongly locally observable by the Hermann–Krener theorem [34].

Observer Design. Under the conditions of Theorem 5.4, it is possible to construct a Luenberger observer explicitly [28], for any number N of elastic coordinates. The observer design procedure is described below (see also the paper [20]).

Lemma 5.6 *Suppose that system (5.60) and (5.70) satisfies the observability condition (5.71) and*

$$0 < \lambda_1 < \lambda_2 < \cdots < \lambda_N, \ b_j d_j > 0, \quad \text{for all } j = 1, 2, \ldots, N.$$

Then, for any control $u(\cdot) \in L^\infty([0, +\infty); \mathbb{R})$ and initial conditions $z(0), \bar{z}(0) \in \mathbb{R}^{2N+2}$, the corresponding solution $z(t)$ of the system of linear approximation for (5.60) tends exponentially to a solution $\bar{z}(t)$ of the following system as $t \to +\infty$:

$$
\begin{aligned}
\dot{\bar{z}}_1 &= (A_{11} - F_1 C_1)\bar{z}_1 + A_{12}\bar{z}_2 + F_1 y_1 + B_1 u, \\
\dot{\bar{z}}_2 &= (A_{22} - F_{22} C_2)\bar{z}_2 + F_{21} y_1 + F_{22} y_2 + B_2 u,
\end{aligned}
\tag{5.73}
$$

where

$$F_1 = (\phi_1, d_0 + \phi_2)^T,$$

$$F_{21} = (0, a_1 - b_1 d_0, \ 0, a_2 - b_2 d_0, \ldots, 0, a_N - b_N d_0)^T,$$

$$F_{22} = (f_1, 0, \ f_2, 0, \ldots, f_N, 0)^T,$$

$$(f_1, f_2, \ldots, f_N)^T = \gamma Q^{-1}(\chi_1, \chi_2, \ldots, \chi_N)^T, \tag{5.74}$$

$$
Q = \begin{pmatrix}
\frac{\lambda_1 d_1}{b_1} + d_1^2 & d_1 d_2 & \cdots & d_1 d_N \\
d_2 d_1 & \frac{\lambda_2 d_2}{b_2} + d_2^2 & \cdots & d_2 d_N \\
\vdots & \vdots & \ddots & \vdots \\
d_N d_1 & d_N d_2 & \cdots & \frac{\lambda_N d_N}{b_N} + d_N^2
\end{pmatrix}.
$$

Here ϕ_1, ϕ_2, and γ are arbitrary positive constants.

Proof We introduce the observational errors: $e_1 = z_1 - \bar{z}_1$ and $e_2 = z_2 - \bar{z}_2$. By subtracting Eq. (5.73) from (5.60) and performing the linearization at the origin, we get the following system:

$$\dot{e}_1 = H_1 e_1 + A_{12} e_2, \ \dot{e}_2 = H_2 e_2,$$

where $H_1 = A_{11} - F_1 C_1$, $H_2 = A_{22} - F_{22} C_2$. Obviously, all roots of the polynomial

$$\det(H_1 - \mu I) = \begin{vmatrix} -\phi_1 - \mu & 1 \\ -\phi_2 & -\mu \end{vmatrix} = \mu^2 + \phi_1 \mu + \phi_2,$$

have negative real parts under the conditions $\phi_1 > 0$ and $\phi_2 > 0$. Let us show that the real parts of the eigenvalues of

$$H_2 = \begin{pmatrix}
-f_1\chi_1 & 1 & \cdots & -f_1\chi_N & 0 \\
-\lambda_1 - b_1 d_1 & 0 & \cdots & -b_1 d_N & 0 \\
-f_2\chi_1 & 0 & \cdots & -f_2\chi_N & 0 \\
-b_2 d_1 & 0 & \cdots & -b_2 d_N & 0 \\
\vdots & \vdots & \ddots & \vdots & \vdots \\
-f_N\chi_1 & 0 & \cdots & -f_N\chi_N & 1 \\
-b_N d_1 & 0 & \cdots & -\lambda_N - b_N d_N & 0
\end{pmatrix}$$

are negative under the conditions of this Lemma. Let us denote

$$e_2 = (\xi_1, \eta_1, \ldots, \xi_N, \eta_N)^T$$

and consider the quadratic form

$$2W(e_2) = \sum_{j=1}^{N} \frac{d_j \eta_j^2}{b_j} + (\xi_1, \xi_2, \ldots, \xi_N) Q (\xi_1, \xi_2, \ldots, \xi_N)^T.$$

It can be shown that the form $2W(e_2)$ is positive definite for $\lambda_j > 0$ and $b_j d_j > 0$. Indeed, all principal minors Δ_j of the matrix Q are positive:

$$\Delta_j = \frac{(\lambda_1 d_1)(\lambda_2 d_2) \cdots (\lambda_j d_j)}{b_1 b_2 \cdots b_j} \left(1 + \sum_{i=1}^{j} \frac{b_i d_i}{\lambda_i} \right) > 0, \quad j = \overline{1, N}.$$

Then the form W is positive definite due to Sylvester's criterion. Inequality $\det(Q) = \Delta_N > 0$ implies also the existence of Q^{-1} in formula (5.74). Computing the time derivative of W along the trajectories of the system $\dot{e}_2 = H_2 e_2$, we get

$$\dot{W}(e_2) = -\gamma (C_2 e_2)^2 \leq 0.$$

Since $\dot{W}$ vanishes on the set $\mathrm{Ker}\, C_2 = \{e_2 \in \mathbb{R}^{2N} \mid C_2 e_2 = 0\}$, we verify the existence of non-trivial trajectories of the system $\dot{e}_2 = H_2 e_2$ on $\mathrm{Ker}\, C_2$. Let $C_2 e_2(t) \equiv 0$, $t \geq 0$, then

$$\frac{d^k}{dt^k} C_2 e_2(t) = C_2 (A_{22} - F_{22} C_2)^k e_2(t) = C_2 A_{22}^k e_2(t) = 0$$

for $t \geq 0$, $k \geq 0$. Hence, for any $t \geq 0$, $e_2(t)$ is a solution of the following system of linear algebraic equations:

$$C_2 A_{22}^k e_2(t) = 0, \quad k = \overline{0, 2N - 1}. \tag{5.75}$$

Under the rank condition (5.71), system (5.75) admits the trivial solution $e_2(t) = 0$ only. Therefore, the linear system $\dot{e}_2 = H_2 e_2$ is asymptotically stable due to the Barbashin–Krasovskii theorem [29].

The matrices H_1 and H_2 are Hurwitz provided that the conditions of this Lemma hold. The error dynamics for the linear approximation of systems (5.60) and (5.73) is described by the following equations:

$$\begin{pmatrix} \dot{e}_1 \\ \dot{e}_2 \end{pmatrix} = \begin{pmatrix} H_1 & A_{12} \\ 0 & H_2 \end{pmatrix} \begin{pmatrix} e_1 \\ e_2 \end{pmatrix}. \tag{5.76}$$

It is easy to see that the spectrum of the matrix of system (5.78) is the union of spectra H_1 and H_2. Therefore, the trivial solution of system (5.78) is exponentially stable, so that $\|z(t) - \bar{z}(t)\| \to 0$ as $t \to +\infty$.

5.7 Observer-Based Stabilization

In this section, we apply a Luenberger-type observer to implement a stabilizing feedback law. The main result in this area is as follows.

Theorem 5.5 *Suppose that system (5.60) and (5.70) satisfies the observability condition (5.71), all λ_j are positive and distinct, $a_j + \lambda_j b_j \neq 0$, and $b_j d_j > 0$ for all $j = \overline{1, N}$. Then the trivial solution $z = 0$, $\bar{z} = 0$ of the extended system (5.60) and*

$$\begin{aligned} \dot{\bar{z}}_1 &= (A_{11} - F_1 C_1)\bar{z}_1 + A_{12}\bar{z}_2 + F_1 y_1 + B_1 u, \\ \dot{\bar{z}}_2 &= (A_{22} - F_{22} C_2)\bar{z}_2 + F_{21} y_1 + F_{22} y_2 + B_2 u, \end{aligned} \tag{5.77}$$

is asymptotically stable in the sense of Lyapunov, where $u = K\bar{z}$ and $y_1 = C_1 z$, $y_2 = C_2 z$. The matrices K, F_1, F_{21}, and F_{22} are as follows:

$$K = (K_1, K_2), \quad K_1 = \left(-d_0 - \frac{h_1 + \sum_{j=1}^{N} a_j(b_j + a_j/\lambda_j)}{h_2}, -\frac{h_0}{h_2} \right),$$

$$K_2 = \left(-d_1 + \frac{a_1 + \lambda_1 b_1}{h_2}, 0, -d_2 + \frac{a_2 + \lambda_2 b_2}{h_2}, 0, \dots, -d_N + \frac{a_N + \lambda_N b_N}{h_2}, 0 \right),$$

$$F_1 = (\phi_1, d_0 + \phi_2)^T, \quad F_{21} = (0, a_1 - b_1 d_0, 0, a_2 - b_2 d_0, \dots, 0, a_N - b_N d_0)^T,$$

$$F_{22} = (f_1, 0, f_2, 0, \dots, f_N, 0)^T, \quad (f_1, f_2, \dots, f_N)^T = \gamma Q^{-1}(\chi_1, \chi_2, \dots, \chi_N)^T,$$

$$
Q = \begin{pmatrix} \frac{\lambda_1 d_1}{b_1} + d_1^2 & d_1 d_2 & \cdots & d_1 d_N \\ d_2 d_1 & \frac{\lambda_2 d_2}{b_2} + d_2^2 & \cdots & d_2 d_N \\ \vdots & \vdots & \ddots & \vdots \\ d_N d_1 & d_N d_2 & \cdots & \frac{\lambda_N d_N}{b_N} + d_N^2 \end{pmatrix}.
$$

Here h_0, h_1, h_2, ϕ_1, ϕ_2, and γ are arbitrary positive constants.

Proof Let us introduce the linear transformation $e_1 = z_1 - \bar{z}_1$, $e_2 = z_2 - \bar{z}_2$ and rewrite extended system (5.60) and (5.77) with respect to variables z, e_1, e_2:

$$
\begin{pmatrix} \dot{z} \\ \dot{e}_1 \\ \dot{e}_2 \end{pmatrix} = \begin{pmatrix} H_0 & -BK_1 & -BK_2 \\ 0 & H_1 & A_{12} \\ 0 & 0 & H_2 \end{pmatrix} \begin{pmatrix} z \\ e_1 \\ e_2 \end{pmatrix} + \begin{pmatrix} R(z, K(z-e)) \\ R_1(z, K(z-e)) \\ R_2(z, K(z-e)) \end{pmatrix}, \tag{5.78}
$$

where

$$
H_0 = \begin{pmatrix} A_{11} + B_1 K_1 & A_{12} + B_1 K_2 \\ A_{21} + B_2 K_1 & A_{22} + B_2 K_2 \end{pmatrix}, \quad H_1 = A_{11} - F_1 C_1, \quad H_2 = A_{22} - F_{22} C_2,
$$

$$
B = (B_1^T, B_2^T)^T, \quad R = (R_1^T, R_2^T)^T.
$$

As it is shown in the proof of Lemma 5.6, the matrices H_1 and H_2 are Hurwitz, i.e. all their eigenvalues have negative real parts. Let us show that the matrix H_0 is Hurwitz as well. For this purpose, we consider an auxiliary system of differential equations

$$
\dot{z} = H_0 z. \tag{5.79}
$$

System (5.79) corresponds to the linear approximation of system (5.60) with the state feedback $u = Kz$ from the proof of Theorem 5.3. The trivial solution of system (5.79) is asymptotically stable due to Theorem 5.3, i.e. the matrix H_0 is Hurwitz.

Since the matrix of the linear part of system (5.78) has a block-triangular form, we conclude that its spectrum is the union of spectra of the Hurwitz matrices H_0, H_1, and H_2. Therefore, the trivial solution of the linear approximation of system (5.78) is asymptotically stable. This fact implies the asymptotic stability of the trivial solution of nonlinear system (5.78) (and also of system (5.60) and (5.77) with $y_1 = C_1 z$, $y_2 = C_2 z$, and $u = K\bar{z}$) due to Lyapunov's theorem on the stability by the linear approximation (see, e.g., [35]).

5.8 Simulation Results

In this section, we compute coefficients of system (5.60) and verify the observability conditions (5.71) in order to justify the applicability of Theorem 5.5 to stabilize a flexible-link manipulator model. The prototype of such a model is the turntable ladder

Table 5.1 Eigenvalues and modal frequencies of oscillations

n	λ_n, c^{-2}	Theoretical frequency v_n, s^{-1}	Measured frequency v_n^*, s^{-1}
1	14.1	0.59	0.6
2	126.5	1.79	1.8
3	354.6	2.99	2.9
4	694.7	4.19	4.25

IVECO DLK 23-12 CS GL described in the paper [36]. The manipulator model has the following mechanical parameters: $l = 25\,$m, $\rho = 64\,$kg/m , $K = 2.3 \times 10^5\,$N, $I_\rho = 2.2 \times 10^4\,$kg m, $EI = 5 \times 10^9\,$N m^2, $m = 100\,$kg, $J_0 = 100\,$kg m^2, $l_0 = 1\,$m, $J_c = 0\,$kg m^2. Consider an equilibrium corresponding to the raising angle $\varphi_0 = \pi/3$.

Table 5.1 shows eigenvalues λ_n ($n = \overline{1,4}$) of the spectral problem (5.18) and (5.19) for the above mechanical parameters. In this table, we also give theoretical frequencies of modal oscillations $v_n = \frac{\sqrt{\lambda_n}}{2\pi}$ and results of experimental measurements of modal frequencies v_n^*.

Thus, it follows from Table 5.1 that the spectrum of (5.18) and (5.19) can be used for an adequate description of the measured modal frequencies.

Let us compute the coefficients of system (5.60) for fixed $N = 2$. For this purpose, we compute eigenfunctions $w_j(x)$ and $\psi_j(x)$ of the problem (5.18) and (5.19). Thus,

$$a_1 = 31.65, \quad a_2 = -7.08, \quad b_1 = 62.95, \quad b_2 = 15.55,$$

$$d_0 = -89.48, \quad d_1 = 307.19, \quad d_2 = 653.39,$$

$$\chi_1 = \psi_1'(l_0) = 0.001, \quad \chi_2 = \psi_2'(l_0) = 0.0014.$$

The above values correspond to the normalization of eigenfunctions with respect to the condition

$$\left\langle \begin{matrix} w_j \\ \psi_j \end{matrix}, \begin{matrix} w_j \\ \psi_j \end{matrix} \right\rangle_{\mathcal{H}} = J_0, \quad j = 1, 2.$$

It can be easily seen that all conditions of Theorem 5.5 are satisfied: $b_1 d_1 > 0$, $b_2 d_2 > 0$,

$$\chi_1 \chi_2 (\lambda_1 - \lambda_2 + b_1 d_1 - b_2 d_2) + b_2 \chi_2^2 d_1 - b_1 \chi_1^2 d_2 = -0.013 \neq 0.$$

To stabilize system (5.60), we apply the control $u = K\bar{z}$ from Theorem 5.5 with the following values of parameters

$$h_0 = \gamma = 10^4, \quad h_1 = 10^3, \quad h_2 = \phi_1 = \phi_2 = 100.$$

Fig. 5.2 Angular position of
the hub B_0

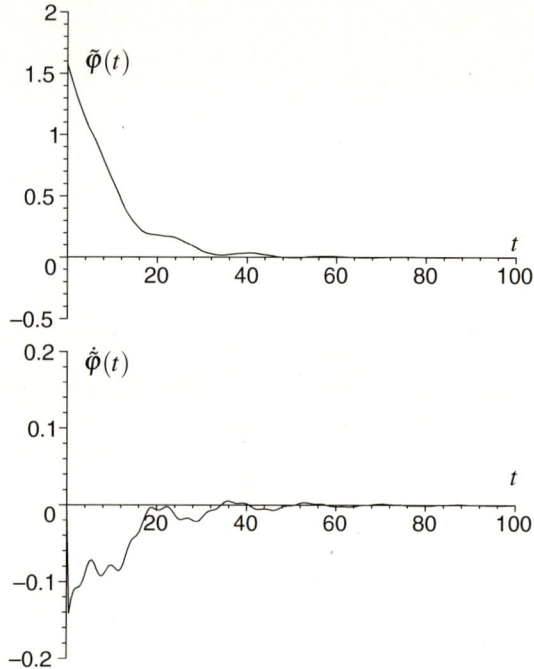

Figures 5.2 and 5.3 show results of the numerical integration of system (5.60) and (5.77) with the following initial conditions:

$$\tilde{\varphi}(0) = \pi/18, \ \dot{\tilde{\varphi}}(0) = 0, \ q_1(0) = q_2(0) = 0, \ \dot{q}_1(0) = \dot{q}_2(0) = 0, \ \tilde{z}(0) = 0.$$
$$(5.80)$$

As we see in Figs. 5.2 and 5.3, the controller proposed is able to steer the system to the equilibrium.

Remark 5.1 In this section, we have not considered optimal stabilization problems or issues of practical implementation of an observer-based control law. Possible recommendations for the choice of constants in control functions from Theorems 5.3 and 5.5 can be obtained from representations of the Lyapunov function (5.63) and its time derivative (5.66) in the proof of Lemma 5.5. Representation $\dot{V} = -h_0 \dot{\tilde{\varphi}}^2$ implies that the constant $h_0 > 0$ has an influence on the decay rate of the Lyapunov function V along solutions of the closed-loop system. On the one hand, the greater value of h_0 implies the better rate of convergence of solutions to the equilibrium (for solutions with $\dot{\tilde{\varphi}} \neq 0$). On the other hand, the coefficient $-\frac{h_0}{h_2}$ with large values of h_0 leads to large values of the control torque for small h_2 in formula (5.68). These arguments suggest the following recommendation: to choose the largest possible value h_0 and, simultaneously, to define h_2 in such a way that the term $-\frac{h_0}{h_2} \dot{\tilde{\varphi}}$ does not exceed the saturation limit for the torque M in the formula $u = Kz$ (for typical values of the angular velocity $\dot{\tilde{\varphi}}$). Then the constant h_1 can be chosen according to

Fig. 5.3 Elastic coordinates of the beam

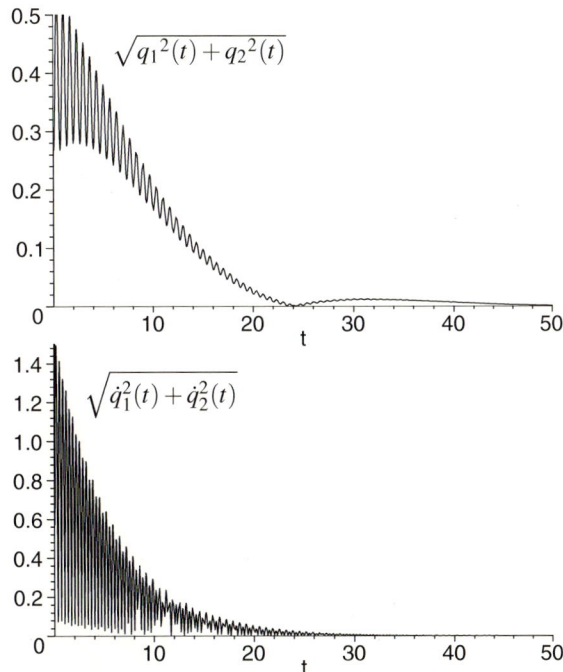

the bounds of $\tilde{\varphi}$ and $\dot{\tilde{\varphi}}$. Indeed, constants h_1 and h_2 define relations between semi-axes of the ellipsoid $V(z) = const$, so using the Lyapunov function $V(z)$, we can estimate the maximal overshoots of $\tilde{\varphi}(t)$ and $\dot{\tilde{\varphi}}(t)$ in terms of h_1 and h_2.

References

1. Timoshenko, S.P.: On the correction for shear of the differential equation for transverse vibrations of prismatic bars. Phil. Mag., **XLI**, 744–746 (1921). Reprinted. In: The Collected Papers of Stephen P. Timoshenko. McGraw-Hill, London (1953)
2. Luo, Z.-H., Guo, B.-Z., Morgul, O.: Stability and Stabilization of Infinite Dimensional Systems with Applications. Springer, London (1999)
3. Kim, J.U., Renardy, Y.: Boundary control of the Timoshenko beam. SIAM J. Control Optim. **25**, 1417–1429 (1987)
4. Krabs, W., Sklyar, G.M.: On the controllability of a slowly rotating Timoshenko beam. J. Anal. Appl. **18**, 437–448 (1999)
5. Krabs, W., Sklyar, G.M.: On the stabilizability of a slowly rotating Timoshenko beam. Z. Anal. Anwend. **19**, 131–145 (2000)
6. Lee, Y.-S., Schultz, W.W.: Eigenvalue analysis of Timoshenko beams and axisymmetric Mindlin plates by the pseudospectral method. J. Sound Vib. **269**, 609–621 (2004)
7. Maxwell, N.D., Asokanthan, S.F.: Modal characteristics of a flexible beam with multiple distributed actuators. J. Sound Vib. **269**, 19–31 (2004)

8. Meurer, T., Zeitz, M.: A modal approach to flatness-based control of flexible structures. PAMM Proc. Appl. Math. Mech. **4**, 133–134 (2004)
9. Morgül, Ö.: Boundary control of a Timoshenko beam attached to a rigid body: planar motion. Int. J. Control **54**, 763–791 (1991)
10. Sklyar, G.M., Szkibiel, G.: Spectral properties of non-homogeneous Timoshenko beam and its rest to rest controllability. J. Math. Anal. Appl. **338**, 1054–1069 (2008)
11. Soufyane, A., Wehbe, A.: Uniform stabilization for the Timoshenko beam by a locally distributed damping. Electron. J. Differ. Equ. **2003**(29), 1–14 (2003)
12. Taylor, S.W.: A smoothing property of a hyperbolic system and boundary controllability. J. Comput. Appl. Math. **114**, 23–40 (2000)
13. Taylor, S.W., Yau, S.C.B.: Boundary control of a rotating Timoshenko beam. Aust. N. Z. Ind. Appl. Math. J. **44**(E), 143–184 (2003)
14. Woittennek, F., Rudolph, J.: Motion planning and boundary control for a rotating Timoshenko beam. PAMM PAMM Proc. Appl. Math. Mech. **2**, 106–107 (2003)
15. Xu, G.Q., Yung, S.P.: Exponential decay rate for a Timoshenko beam with boundary damping. J. Optim. Theory Appl. **123**, 669–693 (2004)
16. Lagnese, J.E., Leugering, G., Schmidt, E.J.P.G.: Modeling, Analysis and Control of Dynamic Elastic Multi-Link Structures. Springer, New York (1994)
17. Balakrishnan, A.V., Shubov, M.A., Peterson, C.A.: Spectral analysis of coupled Euler–Bernoulli and Timoshenko beam model. ZAMM Z. Angew. Math. Mech. **84**(5), 291–313 (2004)
18. Shi, D.-H., Hou, S.H., Feng, D.-X.: Feedback stabilization of a Timoshenko beam with an end mass. Int. J. Control **69**, 285–300 (1998)
19. Tadi, M.: Computational algorithm for controlling a Timoshenko beam. Comput. Methods Appl. Mech. Eng. **153**, 153–165 (1998)
20. Zuyev, A., Sawodny, O.: Stabilization and observability of a rotating Timoshenko beam model. Math. Probl. Eng. Article Id 57238, 1–19 (2007)
21. Berdichevsky, V.: Variational Principles of Continuum Mechanics: I. Fundamentals. Interaction of Mechanics and Mathematics. Springer, New York (2010)
22. Hardy, G., Littlewood, J.E., Pólya, G.: Inequalities, 2nd edn. Cambridge University Press, Cambridge (1952)
23. Geist, B., McLaughlin, J.R.: Double eigenvalues for the uniform Timoshenko beam. Appl. Math. Lett. **10**, 129–134 (1997)
24. Cowper, G.R.: The shear coefficient in Timoshenko's beam theory. Trans. ASME Ser. E. **88**, 335–340 (1966)
25. Kantorovich, L.V., Akilov, G.P.: Functional Analysis. Pergamon Press, Oxford (1982)
26. Donea, J., Huerta, A.: Finite Element Methods for Flow Problems. Wiley, Chichester (2003)
27. Sontag, E.D.: Mathematical Control Theory: Deterministic Finite Dimensional Systems, 2nd edn. Springer, New York (1998)
28. Wonham, W.M.: Linear Multivariable Control: A Geometric Approach, 3rd edn. Springer, New York (1985)
29. Krasovskii, N.N.: Problems of the Theory of Stability of Motion. Stanford University Press, California (1963)
30. Krabs, W.: On Moment Theory and Controllability of One Dimensional Vibrating Systems and Heating Processes. Lecture Notes in Control and Information Sciences, vol. 173. Springer, Berlin (1992)
31. Luo, Z.-H., Guo, B.-Z.: Further theoretical results on direct strain feedback control of flexible robot arms. IEEE Trans. Autom. Control **40**, 747–751 (1995)
32. Luo, Z.-H., Guo, B.-Z.: Shear force feedback control of a single-link flexible robot with a revolute joint. IEEE Trans. Autom. Control **42**, 53–65 (1997)
33. Komornik, V., Loreti, P.: Fourier Series in Control Theory. Springer, New York (2005)
34. Hermann, R., Krener, A.J.: Nonlinear controllability and observability. IEEE Trans. Autom. Control **22**, 728–740 (1977)

35. Lyapunov, A.M.: The General Problem of the Stability of Motion. (A. T. Fuller trans.) Taylor & Francis, London (1992)
36. Sawodny, O., Aschemann, H., Bulach, A.: Mechatronical designed control of fire-rescue turntable ladders as flexible link robots, Preprints of the 15th IFAC World Congress, Barcelona. CD-ROM file 385.pdf (2002)

Chapter 6
Control and Stabilization of a Rotating Kirchhoff Plate

Abstract In this chapter, we study a mechanical system consisting of a rotating rigid body and the Kirchhoff plate. We derive the equations of motion of this system with modal coordinates by assuming that the control is the angular acceleration of the body. The linearized control system is shown to be neither controllable nor stabilizable in general case. We propose a state feedback control that ensures partial asymptotic stability of this system. An estimate of the reachable set is proposed by using the approach of Chap. 4. An infinite-dimensional subsystem corresponding to the modes with odd indices is shown to be approximately controllable under an additional assumption that the ratio of the sides of the rectangular plate is an irrational algebraic number.

In the monograph [1], five different mathematical models of thin plates have been considered: the Kirchhoff model; the Mindlin–Timoshenko model; the von Karman model; a viscoelastic plate with long-range memory; and a thermoelastic plate. The stabilization problem has been solved in [1] for dynamical plate models with boundary inputs, and the exact boundary controllability problem is treated in [2]. Systems of linked Kirchhoff and Reissner plates have been considered in the monograph [3]. Controllability conditions have been obtained in [3] for transmission problems for thin plates.

The monograph [4] deals with boundary or point control problems for systems of coupled partial differential equations. Several models of flexible structures such as structurally damped beams, plates, and shells are considered. It is emphasized that, when structural damping is nonexistent, one has to introduce other stabilization schemes, in particular, boundary dampers. A part of that work focuses on the design of boundary feedbacks which provide uniform stability properties of the structurally damped plate and of the corresponding structural acoustic models. Other results on the dynamics and control of thin plates are available in papers [5–11].

In the monograph [12, Chap. 4], a variational form of the equations of motion of a thin plate with distributed piezo-actuators is studied. The existence and uniqueness properties of solutions are established there by using the Lax–Milgram theorem. A thin plate example with distributed MFC (macrofiber composite) piezo-actuators is considered as a prototypical smart structure.

© Springer International Publishing Switzerland 2015   169
A.L. Zuyev, *Partial Stabilization and Control of Distributed Parameter Systems with Elastic Elements*, Lecture Notes in Control and Information Sciences 458,
DOI 10.1007/978-3-319-11532-0_6

Another important model of plates with large deflections is described by the von Karman equations. The monograph [13] presents recent mathematical results concerning the well-posedness and regularity of solutions as well as their qualitative behavior as $t \to +\infty$. This includes the localization of attractors for the von Karman model with damping and the study of inertial manifolds.

As it was mentioned in the monograph [1], the issue of stabilizing the motion of composite structures (such as body-beam-plate systems) has yet to be systematically studied from a distributed parameter systems point of view. In this chapter, we consider a flexible structure consisting of a rotating rigid body and the Kirchhoff plate. Our study extends the approach of the paper [14] for a model with inertial terms and two-dimensional input.

6.1 Rotating Kirchhoff Plate

Consider a mechanical system that consists of a flexible plate attached on its boundary to a rotating rigid body B (Fig. 6.1).

Let $Ox_1x_2x_3$ be a Cartesian frame associated with B. Suppose that the plate in its undeformed state occupies the following closed domain:

$$(x_1, x_2) \in \Omega = [0, l_1] \times [0, l_2], \quad |x_3| \le h/2,$$

where $h > 0$ is the thickness of the plate. Suppose also that the median surface of the plate is described by the equation $x_3 = w(x_1, x_2, t)$ at each time $t \ge 0$.

We use the Kirchhoff plate model (see, e.g., [2, 15]) which is based on the following fourth order partial differential equation:

$$\rho h \frac{\partial^2 w(x_1, x_2, t)}{\partial t^2} - I_\rho \Delta \frac{\partial^2 w(x_1, x_2, t)}{\partial t^2} + D\Delta^2 w(x_1, x_2, t) = F, \quad (x_1, x_2) \in \Omega.$$

$$(6.1)$$

Fig. 6.1 Rigid body with the Kirchhoff plate

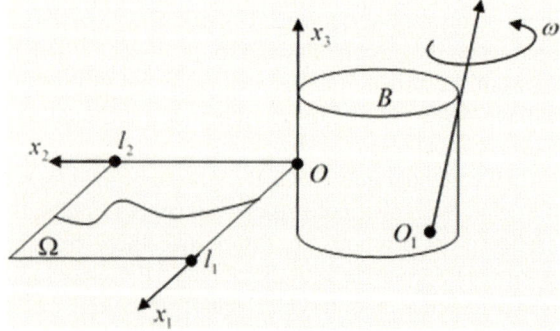

Here $\Delta = \frac{\partial^2}{\partial x_1^2} + \frac{\partial^2}{\partial x_2^2}$ is the Laplace operator, $\rho > 0$ is the density (mass per unit volume of the plate), $D > 0$ is the flexural rigidity, and F is the transverse component of the force acting on the plate. The coefficient I_ρ in Eq. (6.1) is the polar moment of inertia of an element of the plate; for the plate with constant density it is given by the formula

$$I_\rho = \frac{\rho h^3}{12}.$$

For a simplified Kirchhoff model, this coefficient is usually neglected ($I_\rho = 0$), see, e.g., [15].

We assume that the plate is hinged at the boundary of the domain Ω, i.e. the displacement vector and the bending moment vanish on $\partial\Omega$:

$$w(x_1, x_2, t)|_{\partial\Omega} = 0, \tag{6.2}$$

$$\frac{\partial^2 w(x_1, x_2, t)}{\partial x_1^2} + \nu \frac{\partial^2 w(x_1, x_2, t)}{\partial x_2^2}\Bigg|_{x_1=0,\, x_1=l_1} = 0, \tag{6.3}$$

$$\frac{\partial^2 w(x_1, x_2, t)}{\partial x_2^2} + \nu \frac{\partial^2 w(x_1, x_2, t)}{\partial x_1^2}\Bigg|_{x_2=0,\, x_2=l_2} = 0, \tag{6.4}$$

where ν is Poisson's ratio.

The right-hand side of Eq. (6.1) contains the inertia force term F due to the motion of the rigid body B. We denote the angular velocity vector of B by $\omega = (\omega_1, \omega_2, \omega_3)$ and assume that the body has a fixed point O_1 with coordinates (a_1, a_2, a_3). All the coordinates ω_i and a_i are evaluated in the Cartesian frame $Ox_1x_2x_3$. By using these notations, we compute the inertia force:

$$F = -\rho h[(x_1 - a_1)(\omega_1\omega_3 - \dot\omega_2) + (x_2 - a_2)(\omega_2\omega_3 + \dot\omega_1)$$
$$+ (\omega_1^2 + \omega_2^2)(a_3 - w(x_1, x_2, t)) + \ddot w(x_1, x_2, t)]. \tag{6.5}$$

Let us rewrite Eq. (6.1) with the force (6.5):

$$\frac{\partial^2 w(x_1, x_2, t)}{\partial t^2} - \frac{I_\rho}{2\rho}\Delta\frac{\partial^2 w(x_1, x_2, t)}{\partial t^2} + \alpha^2\Delta^2 w(x_1, x_2, t)$$
$$= -\frac{1}{2}[(x_1 - a_1)(\omega_1\omega_3 - \dot\omega_2) + (x_2 - a_2)(\omega_2\omega_3 + \dot\omega_1) + (\omega_1^2 + \omega_2^2)$$
$$\times (a_3 - w(x_1, x_2, t))] = f(x_1, x_2, t), \tag{6.6}$$

where $\alpha^2 = \dfrac{D}{2\rho h} > 0$. In the sequel, we assume that the mechanical parameters $\rho > 0$, $D \geq 0$, and $\alpha \geq 0$ are constant.

Thus, we have obtained the mathematical model (6.2)–(6.4), (6.6) of vibrations of the Kirchhoff plate attached to a rotating body on the boundary of the domain Ω. Note that a particular case of the rotation around a fixed axis has been considered in the paper [14].

Separation of variables. To study properties of the model, we follow the Fourier method and substitute

$$w(x_1, x_2, t) = X(x)q(t), \qquad x = (x_1, x_2)$$

into the boundary value problem (6.2)–(6.4), (6.6) with $f = 0$. As a result of the separation of variables, we get

$$\frac{\ddot{q}}{q} = -\frac{\alpha^2 \Delta^2 X}{X - \frac{I_\rho}{2\rho}\Delta X} = -\lambda = \text{const.}$$

Consider the equation

$$\alpha^2 \Delta^2 X = \lambda\left(X - \frac{I_\rho}{2\rho}\Delta X\right). \tag{6.7}$$

We suppose that $X(x_1, x_2) = X_1(x_1)X_2(x_2)$ and denote

$$\frac{X_1''}{X_1} = \mu_1, \qquad \frac{X_2''}{X_2} = \mu_2, \qquad \mu = \mu_1 + \mu_2.$$

Then $\Delta X = (\mu_1 + \mu_2)X_1 X_2 = \mu X$. By substituting this expression into Eq. (6.7), we get $\lambda = \dfrac{\alpha^2(\mu_1 + \mu_2)^2}{1 - \frac{I_\rho}{2\rho}(\mu_1 + \mu_2)}$. Equation (6.6) takes the form

$$\ddot{q}(t) + \lambda q(t) = 0,$$

where μ_1 and μ_2 are eigenvalues of the following spectral problems:

$$\begin{cases} X_1''(x_1) = \mu_1 X_1(x_1), \\ X_1(0) = X_1(l_1) = 0, \end{cases} \qquad 0 \leq x_1 \leq l_1, \tag{6.8}$$

$$\begin{cases} X_2''(x_2) = \mu_2 X_2(x_2), \\ X_2(0) = X_2(l_2) = 0, \end{cases} \qquad 0 \leq x_2 \leq l_2. \tag{6.9}$$

It is a well-known fact that the spectrum of each problem (6.8) and (6.10) is discrete (see, e.g., [16]):

$$\mu_1 = \mu_{1k}, \quad \mu_2 = \mu_{2j}, \quad k, j \in \mathbb{N},$$

where

$$\mu_{1k} = -\left(\frac{\pi k}{l_1}\right)^2, \quad \mu_{2j} = -\left(\frac{\pi j}{l_2}\right)^2. \tag{6.10}$$

Eigenvalues of the spectral problems (6.8) and (6.10) are related to eigenfunctions $\{X_{1k}\}_{k=1}^{\infty}$ and $\{X_{2j}\}_{j=1}^{\infty}$. We normalize these functions so that they form orthonormal bases in $L^2(0, l_1)$ and $L^2(0, l_2)$, respectively:

$$X_{1k}(x_1) = \sqrt{\frac{2}{l_1}} \sin\left(\frac{\pi k x_1}{l_1}\right), \quad X_{2j}(x_2) = \sqrt{\frac{2}{l_2}} \sin\left(\frac{\pi j x_2}{l_2}\right). \tag{6.11}$$

Let us find a formal solution of the boundary value problem (6.6), (6.2)–(6.4) as the Fourier series

$$w(x_1, x_2, t) = \sum_{k,j=1}^{\infty} C_{kj}(t) X_{1k}(x_1) X_{2j}(x_2).$$

Differentiating this series term-by-term and substituting into Eq. (6.6) with the use of relations $X_{1k}'' = \mu_{1k} X_{1k}$ and $X_{2j}'' = \mu_{2j} X_{2j}$, we get

$$\sum_{k,j=1}^{\infty} \ddot{C}_{kj} X_{1k} X_{2j} - \frac{I_\rho}{2\rho} \sum_{k,j=1}^{\infty} \ddot{C}_{kj} (\mu_{1k} X_{1k} X_{2j} + \mu_{2j} X_{1k} X_{2j})$$

$$+ \alpha^2 \sum_{k,j=1}^{\infty} C_{kj} \times (\mu_{1k}^2 X_{1k} X_{2j} + 2\mu_{1k}\mu_{2j} X_{1k} X_{2j} + \mu_{2j}^2 X_{1k} X_{2j})$$

$$= \sum_{k,j=1}^{\infty} f_{kj} X_{1k} X_{2j}. \tag{6.12}$$

The above transformations lead to the following infinite system of ordinary differential equations:

$$\ddot{C}_{kj} + \lambda_{kj} C_{kj} = f_{kj}, \quad \lambda_{kj} = \frac{\alpha^2 (\mu_{1k} + \mu_{2j})^2}{1 - \frac{I_\rho}{2\rho}(\mu_{1k} + \mu_{2j})}, \quad (k, j) \in \mathbb{N}^2, \tag{6.13}$$

where f_{kj} are the Fourier coefficients of the right-hand side of Eq. (6.6) with respect to the orthonormal system $\{X_{1k}(x_1) X_{2j}(x_2)\}_{k,j=1}^{\infty}$ in $L^2(\Omega)$:

$$f_{kj} = \frac{2}{\left(1 + \frac{I_\rho}{2\rho}\left(\frac{(\pi k l_2)^2 + (\pi j l_1)^2}{(l_1 l_2)^2}\right)\right)\sqrt{l_1 l_2}}$$

$$\times \int_{\Omega} f(x_1, x_2, t) \sin\left(\frac{\pi k x_1}{l_1}\right) \sin\left(\frac{\pi j x_2}{l_2}\right) dx_1 dx_2$$

$$= \frac{\sqrt{l_1 l_2}}{\pi^2 k j \left(1 + \frac{I_\rho}{2\rho}\left(\frac{(\pi k l_2)^2 + (\pi j l_1)^2}{(l_1 l_2)^2}\right)\right)} \left(-\omega_1 \omega_3 l_1 (-1)^k \left((-1)^j - 1\right)\right.$$

$$+ \dot{\omega}_2 l_1 (-1)^k \left((-1)^j - 1\right) - a_1 \dot{\omega}_2 ((-1)^k - 1)\left((-1)^j - 1\right)$$

$$+ a_1 \omega_1 \omega_3 ((-1)^k - 1)((-1)^j - 1) - \omega_2 \omega_3 l_2 \left((-1)^k - 1\right)(-1)^j$$

$$- \dot{\omega}_1 l_2 \left((-1)^k - 1\right)(-1)^j + a_2 \omega_2 \omega_3 \left((-1)^k - 1\right)$$

$$\times \left((-1)^j - 1\right) + a_2 \dot{\omega}_1 \left((-1)^k - 1\right)\left((-1)^j - 1\right)$$

$$\left. - a_3 (\omega_1^2 + \omega_2^2)\left((-1)^k - 1\right)\left((-1)^j - 1\right) + \frac{(\omega_1^2 + \omega_2^2)}{\sqrt{l_1 l_2}} C_{kj}(t).$$

In order to study the influence of the carrier body B on the plate vibrations, we will treat $u_1(t) = \dot{\omega}_1(t)$ and $u_2(t) = \dot{\omega}_2(t)$ and controls. Moreover, to simplify the model, we assume that the rotation of B is slow and the deformation of the plate is small, so that we will consider the linear approximation of system (6.13) in the sequel[1]:

$$\ddot{C}_{kj} + C_{kj} \lambda_{kj} = b_{kj}^1 u_1(t) + b_{kj}^2 u_2(t), \quad (k, j) \in \mathbb{N}^2, \tag{6.14}$$

where

$$\frac{\pi^2 k j \left(1 + \frac{I_\rho}{2\rho}\left(\frac{(\pi k l_2)^2 + (\pi j l_1)^2}{(l_1 l_2)^2}\right)\right)}{2\sqrt{l_1 l_2}} b_{kj}^1 = \begin{cases} 0, & \text{for even } k, \\ l_2, & \text{for odd } k \text{ and even } j, \\ 2a_2 - l_2, & \text{for odd } k \text{ and odd } j, \end{cases}$$

$$\frac{\pi^2 k j \left(1 + \frac{I_\rho}{2\rho}\left(\frac{(\pi k l_2)^2 + (\pi j l_1)^2}{(l_1 l_2)^2}\right)\right)}{2\sqrt{l_1 l_2}} b_{kj}^2 = \begin{cases} 0, & \text{for even } j, \\ -l_1, & \text{for even } k \text{ and odd } j, \\ l_1 - 2a_1, & \text{for odd } k \text{ and odd } j. \end{cases}$$

We introduce the following change of variables in system (6.14):

$$\sqrt{\lambda_{kj}} C_{kj} = P_{kj}(t), \qquad \dot{C}_{kj}(t) = Q_{kj}(t).$$

---

[1] We neglect the terms of order $o(|\omega_k|, |\dot{\omega}_k|, |C_{kj}|, |\dot{C}_{kj}|)$.

Then system (6.14) takes the form

$$\frac{d}{dt}\begin{pmatrix} P_{kj}(t) \\ Q_{kj}(t) \end{pmatrix} = \begin{pmatrix} 0 & \sqrt{\lambda_{kj}} \\ -\sqrt{\lambda_{kj}} & 0 \end{pmatrix} \begin{pmatrix} P_{kj}(t) \\ Q_{kj}(t) \end{pmatrix} + \begin{pmatrix} 0 & 0 \\ b_{kj}^1 & b_{kj}^2 \end{pmatrix} \begin{pmatrix} u_1(t) \\ u_2(t) \end{pmatrix}, \quad (k, j) \in \mathbb{N}^2.$$

$$(6.15)$$

System (6.15) is considered as a mathematical model of the rotating Kirchhoff plate with modal coordinates P_{kj} and Q_{kj}.

6.2 Partial Stabilization of the Equilibrium

To simplify further analysis, we will consider the "standard" Kirchhoff plate with $I_\rho = 0$ in the sequel. Then the coefficients of Eq. (6.15) are represented through mechanical parameters as follows:

$$\lambda_{kj} = \alpha^2 \left(\left(\frac{\pi k}{l_1} \right)^2 + \left(\frac{\pi j}{l_2} \right)^2 \right)^2, \qquad (6.16)$$

$$b_{kj}^1 = \begin{cases} 0, & \text{for even } k, \\ \dfrac{2\sqrt{l_1 l_2^3}}{\pi^2 kj}, & \text{for odd } k \text{ and even } j, \\ \dfrac{2\sqrt{l_1 l_2}(2a_2 - l_2)}{\pi^2 kj}, & \text{for odd } k \text{ and odd } j, \end{cases}$$

$$b_{kj}^2 = \begin{cases} -\dfrac{2\sqrt{l_1^3 l_2}}{\pi^2 kj}, & \text{for even } k \text{ and odd } j, \\ 0, & \text{for even } j, \\ \dfrac{2\sqrt{l_1 l_2}(l_1 - 2a_1)}{\pi^2 kj}, & \text{for odd } k \text{ and odd } j. \end{cases}$$

Let $n \mapsto (k_n, j_n)$ be a one-to-one correspondence between the sets $\mathbb{N}$ and $\mathbb{N}^2$. In Sects. 6.2 and 6.3, we use notations with a single index n instead of the corresponding double index notations with (k_n, j_n) as follows:

$$\xi_n := P_{kj}, \quad \eta_n := Q_{kj}, \quad \beta_n := \sqrt{\lambda_{kj}} > 0, \quad \varphi_n := b_{kj}^1,$$
$$\psi_n := b_{kj}^2 \quad \text{for } (k, j) = (k_n, j_n). \qquad (6.17)$$

Thus, we represent system (6.15) in the operator form as follows:

$$\dot{x} = Ax + Bu, \quad x \in \ell^2, \quad u \in \mathbb{R}^2, \qquad (6.18)$$

where

$$x = \begin{pmatrix} \xi_1 \\ \eta_1 \\ \xi_2 \\ \eta_2 \\ \vdots \end{pmatrix}, \quad A = \begin{pmatrix} A_1 & 0 & \cdots \\ 0 & A_2 & \cdots \\ \vdots & \vdots & \ddots \end{pmatrix}, \quad B = \begin{pmatrix} B_1 \\ B_2 \\ \vdots \end{pmatrix}, \quad u = \begin{pmatrix} u_1 \\ u_2 \end{pmatrix},$$

$$A_n = \begin{pmatrix} 0 & \beta_n \\ -\beta_n & 0 \end{pmatrix}, \quad B_n = \begin{pmatrix} 0 & 0 \\ \varphi_n & \psi_n \end{pmatrix}, \quad n = 1, 2, \ldots \tag{6.19}$$

We use the standard inner product of vectors $x = (\xi_1, \eta_1, \xi_2, \eta_2, \ldots)^T$ and $\bar{x} = (\bar{\xi}_1, \bar{\eta}_1, \bar{\xi}_2, \bar{\eta}_2, \ldots)^T$ in the Hilbert space ℓ^2:

$$\langle x, \bar{x} \rangle_{\ell^2} = \sum_{n=1}^{\infty} \left(\xi_n \bar{\xi}_n + \eta_n \bar{\eta}_n \right).$$

Let us find a feedback control $u = v(x)$ that ensures the partial asymptotic stability of the trivial solution of system (6.18). For this purpose, we consider the following Lyapunov functional $V : \ell^2 \to \mathbb{R}$:

$$V(x) = \sum_{\substack{n=1 \\ n \notin S}}^{\infty} \left(\xi_n^2 + \eta_n^2 \right), \tag{6.20}$$

where

$$S = \{ n \in \mathbb{N} \mid \varphi_n = \psi_n = 0 \}.$$

The time derivative of the functional V along the trajectories of system (6.18) is

$$\dot{V} = 2 \sum_{\substack{n=1 \\ n \notin S}}^{\infty} \left(\xi_n \dot{\xi}_n + \eta_n \dot{\eta}_n \right) = 2u_1 \left(\sum_{\substack{n=1 \\ n \notin S}}^{\infty} \eta_n \varphi_n \right) + 2u_2 \left(\sum_{\substack{n=1 \\ n \notin S}}^{\infty} \eta_n \psi_n \right).$$

Let us define the following feedback controls:

$$v_1(x) = -\frac{k_1}{2} \sum_{\substack{n=1 \\ n \notin S}}^{\infty} \varphi_n \eta_n, \quad v_2(x) = -\frac{k_2}{2} \sum_{\substack{n=1 \\ n \notin S}}^{\infty} \psi_n \eta_n, \quad (k_1 \geq 0, \; k_2 \geq 0). \tag{6.21}$$

Then the time derivative of the functional V along the trajectories of the closed-loop system (6.18) with $u_1 = v_1(x)$, $u_2 = v_2(x)$ reads as

$$\dot{V} = -k_1 \left(\sum_{\substack{n=1 \\ n \notin S}}^{\infty} \varphi_n \eta_n \right)^2 - k_2 \left(\sum_{\substack{n=1 \\ n \notin S}}^{\infty} \psi_n \eta_n \right)^2 \leq 0.$$

The closed-loop system (6.18) with $u = v(x)$ is represented in the operator form as

$$\dot{x} = Fx, \quad x(0) = x_0 \in \ell^2, \tag{6.22}$$

where $F = A + BK$ with $K = -\dfrac{1}{2} \begin{pmatrix} 0 & k_1\varphi_1 & 0 & k_1\varphi_2 & \cdots \\ 0 & k_2\psi_1 & 0 & k_2\psi_2 & \cdots \end{pmatrix}$.

The operator F is the infinitesimal generator of a C_0-semigroup of bounded linear operators $\{e^{tF}\}_{t \geq 0}$ on ℓ^2 due to the Lumer–Phillips theorem (Theorem 1.3).

Consider a continuous functional

$$\mu(x) = \sum_{\substack{n=1 \\ n \notin S}}^{\infty} \left(\xi_n^2 + \eta_n^2 \right) \tag{6.23}$$

in the space ℓ^2.

The main result of this section is the following theorem on the partial asymptotic stability in the sense of Definition 2.5.

Theorem 6.1 *Let k_1 and k_2 be arbitrary positive constants. Then the trivial solution of the closed-loop system (6.18) with the feedback controls $u_1 = v_1(x)$ and $u_2 = v_2(x)$ of form (6.21) is asymptotically stable with respect to the functional μ of form (6.23).*

Proof The proof is based on the application of Theorem 2.4. Condition (i) of Theorem 2.4 is satisfied because

$$\mu(x) \leq V(x) = \|x\|^2, \quad \forall x \in \ell^2.$$

Condition (ii) follows from the fact that

$$\dot{V} = -k_1 \left(\sum_{\substack{n=1 \\ n \notin S}}^{\infty} \varphi_n \eta_n \right)^2 - k_2 \left(\sum_{\substack{n=1 \\ n \notin S}}^{\infty} \psi_n \eta_n \right)^2 \leq 0$$

as $k_1 > 0$ and $k_2 > 0$, for all $x \in D(A)$, where

$$D(A) = \{x \in \ell^2 \mid \sum_{\substack{n=1 \\ n \notin S}}^{\infty} \beta_n^2 \left(\xi_n^2 + \eta_n^2 \right) < \infty\}.$$

To verify condition (iii) of Theorem 2.4 we prove the precompactness of the trajectories of linear differential equation (6.22) by using the Dafermos–Slemrod Theorem (Theorem 1.6). For this purpose, we prove the compactness of the resolvent $(\lambda F + I)^{-1} : \ell^2 \to \ell^2$ for some $\lambda > 0$.

Let us consider the equation $Ix + \lambda Ax + \lambda Bu = \bar{x}$ with respect to x, where $\lambda = \text{const}$. Its component form is

$$(I + \lambda A) \begin{pmatrix} \xi_n \\ \eta_n \end{pmatrix} = \begin{pmatrix} \bar{\xi}_n \\ \bar{\eta}_n - \lambda \varphi_n u_1 - \lambda \psi_n u_2 \end{pmatrix}.$$

Solving the above equation by the inverse matrix method, we get

$$\begin{pmatrix} \xi_n \\ \eta_n \end{pmatrix} = \frac{1}{1 + (\lambda \beta_n)^2} \begin{pmatrix} 1 & -\lambda \beta_n \\ \lambda \beta_n & 1 \end{pmatrix} \begin{pmatrix} \bar{\xi}_n \\ \bar{\eta}_n - \lambda \varphi_n u_1 - \lambda \psi_n u_2 \end{pmatrix}. \tag{6.24}$$

Let us substitute (6.24) into the expression for $u = v(x)$ of form (6.21):

$$\begin{pmatrix} v_1(\bar{x}) \\ v_2(\bar{x}) \end{pmatrix} = - \begin{pmatrix} \dfrac{k_1}{2} \displaystyle\sum_{\substack{n=1 \\ n \notin S}}^{\infty} \dfrac{\varphi_n}{1 + (\lambda \beta_n)^2} \left(\lambda \beta_n \bar{\xi}_n + \bar{\eta}_n - \lambda \varphi_n u_1 - \lambda \psi_n u_2 \right) \\ \dfrac{k_2}{2} \displaystyle\sum_{\substack{n=1 \\ n \notin S}}^{\infty} \dfrac{\psi_n}{1 + (\lambda \beta_n)^2} \left(\lambda \beta_n \bar{\xi}_n + \bar{\eta}_n - \lambda \varphi_n u_1 - \lambda \psi_n u_2 \right) \end{pmatrix}. \tag{6.25}$$

We transform (6.25) as follows:

$$\begin{cases} v_1 \left(1 - \dfrac{k_1}{2} \displaystyle\sum_{\substack{n=1 \\ n \notin S}}^{\infty} \dfrac{\lambda \varphi_n^2}{1 + (\lambda \beta_n)^2} \right) - v_2 \left(\dfrac{k_1}{2} \displaystyle\sum_{\substack{n=1 \\ n \notin S}}^{\infty} \dfrac{\lambda \varphi_n \psi_n}{1 + (\lambda \beta_n)^2} \right) = - \dfrac{k_1}{2} \displaystyle\sum_{\substack{n=1 \\ n \notin S}}^{\infty} \dfrac{\varphi_n \left(\lambda \beta_n \bar{\xi}_n + \bar{\eta}_n \right)}{1 + (\lambda \beta_n)^2}, \\[6mm] v_1 \left(- \dfrac{k_2}{2} \displaystyle\sum_{\substack{n=1 \\ n \notin S}}^{\infty} \dfrac{\lambda \varphi_n \psi_n}{1 + (\lambda \beta_n)^2} \right) + v_2 \left(1 - \dfrac{k_2}{2} \displaystyle\sum_{\substack{n=1 \\ n \notin S}}^{\infty} \dfrac{\lambda \psi_n^2}{1 + (\lambda \beta_n)^2} \right) = - \dfrac{k_2}{2} \displaystyle\sum_{\substack{n=1 \\ n \notin S}}^{\infty} \dfrac{\psi_n \left(\lambda \beta_n \bar{\xi}_n + \bar{\eta}_n \right)}{1 + (\lambda \beta_n)^2}. \end{cases} \tag{6.26}$$

Then we express the functionals $v_1(\bar{x})$ and $v_2(\bar{x})$ from (6.26):

$$v_1(\bar{x}) = \frac{\Delta_1}{\Delta}, \qquad v_2(\bar{x}) = \frac{\Delta_2}{\Delta}, \tag{6.27}$$

where

$$\Delta = 1 - \frac{k_2}{2} \sum_{\substack{n=1 \\ n \notin S}}^{\infty} \frac{\lambda \psi_n^2}{1 + (\lambda \beta_n)^2} - \frac{k_1}{2} \sum_{\substack{n=1 \\ n \notin S}}^{\infty} \frac{\lambda \varphi_n^2}{1 + (\lambda \beta_n)^2} + \frac{k_1 k_2}{4} \sum_{\substack{n=1 \\ n \notin S}}^{\infty} \frac{\lambda \varphi_n^2}{1 + (\lambda \beta_n)^2}$$

$$\times \sum_{\substack{n=1 \\ n \notin S}}^{\infty} \frac{\lambda \psi_n^2}{1 + (\lambda \beta_n)^2} - \frac{k_1 k_2}{4} \left(\sum_{\substack{n=1 \\ n \notin S}}^{\infty} \frac{\lambda \varphi_n \psi_n}{1 + (\lambda \beta_n)^2} \right)^2,$$

$$\Delta_1 = -\frac{k_1}{2} \sum_{\substack{n=1 \\ n \notin S}}^{\infty} \frac{\varphi_n \left(\lambda \beta_n \overline{\xi}_n + \overline{\eta}_n \right)}{1 + (\lambda \beta_n)^2} + \frac{k_1 k_2}{4} \sum_{\substack{n=1 \\ n \notin S}}^{\infty} \frac{\varphi_n \left(\lambda \beta_n \overline{\xi}_n + \overline{\eta}_n \right)}{1 + (\lambda \beta_n)^2} \sum_{\substack{n=1 \\ n \notin S}}^{\infty} \frac{\lambda \psi_n^2}{1 + (\lambda \beta_n)^2}$$

$$- \frac{k_1 k_2}{4} \sum_{\substack{n=1 \\ n \notin S}}^{\infty} \frac{\psi_n \left(\lambda \beta_n \overline{\xi}_n + \overline{\eta}_n \right)}{1 + (\lambda \beta_n)^2} \sum_{\substack{n=1 \\ n \notin S}}^{\infty} \frac{\lambda \varphi_n \psi_n}{1 + (\lambda \beta_n)^2},$$

$$\Delta_2 = -\frac{k_2}{2} \sum_{\substack{n=1 \\ n \notin S}}^{\infty} \frac{\psi_n \left(\lambda \beta_n \overline{\xi}_n + \overline{\eta}_n \right)}{1 + (\lambda \beta_n)^2} + \frac{k_1 k_2}{4} \sum_{\substack{n=1 \\ n \notin S}}^{\infty} \frac{\psi_n \left(\lambda \beta_n \overline{\xi}_n + \overline{\eta}_n \right)}{1 + (\lambda \beta_n)^2} \sum_{\substack{n=1 \\ n \notin S}}^{\infty} \frac{\lambda \varphi_n^2}{1 + (\lambda \beta_n)^2}$$

$$- \frac{k_1 k_2}{4} \sum_{\substack{n=1 \\ n \notin S}}^{\infty} \frac{\varphi_n \left(\lambda \beta_n \overline{\xi}_n + \overline{\eta}_n \right)}{1 + (\lambda \beta_n)^2} \sum_{\substack{n=1 \\ n \notin S}}^{\infty} \frac{\lambda \varphi_n \psi_n}{1 + (\lambda \beta_n)^2}.$$

Formulas (6.27) define the linear functional $v(\overline{x})$ in ℓ^2. Let us estimate $v_1(\overline{x})$ and $v_2(\overline{x})$ in order to show the boundedness of $v(\overline{x})$ for an arbitrary $\lambda > 0$. Since $|v_1(\overline{x})| = \left| \frac{\Delta_1}{\Delta} \right| = \frac{|\Delta_1|}{|\Delta|}$ and $|v_2(\overline{x})| = \left| \frac{\Delta_2}{\Delta} \right| = \frac{|\Delta_2|}{|\Delta|}$, we estimate the numerator and denominator separately as follows:

$$|\Delta_1| \leq \frac{k_1}{2} \sum_{\substack{n=1 \\ n \notin S}}^{\infty} \frac{|\varphi_n| |\lambda \beta_n \overline{\xi}_n|}{1 + (\lambda \beta_n)^2} + \frac{k_1}{2} \sum_{\substack{n=1 \\ n \notin S}}^{\infty} \frac{|\varphi_n| |\overline{\eta}_n|}{1 + (\lambda \beta_n)^2} + \frac{k_1 k_2}{4} \sum_{\substack{n=1 \\ n \notin S}}^{\infty} \frac{\lambda \psi_n^2}{1 + (\lambda \beta_n)^2}$$

$$\times \sum_{\substack{n=1 \\ n \notin S}}^{\infty} \frac{|\varphi_n| |\lambda \beta_n \overline{\xi}_n|}{1 + (\lambda \beta_n)^2} + \frac{k_1 k_2}{4} \sum_{\substack{n=1 \\ n \notin S}}^{\infty} \frac{\lambda \psi_n^2}{1 + (\lambda \beta_n)^2} \sum_{\substack{n=1 \\ n \notin S}}^{\infty} \frac{|\varphi_n| |\overline{\eta}_n|}{1 + (\lambda \beta_n)^2} + \frac{k_1 k_2}{4}$$

$$\times \sum_{\substack{n=1 \\ n \notin S}}^{\infty} \frac{\lambda |\varphi_n \psi_n|}{1 + (\lambda \beta_n)^2} \sum_{\substack{n=1 \\ n \notin S}}^{\infty} \frac{|\psi_n| |\lambda \beta_n \overline{\xi}_n|}{1 + (\lambda \beta_n)^2} + \frac{k_1 k_2}{4} \sum_{\substack{n=1 \\ n \notin S}}^{\infty} \frac{\lambda |\varphi_n \psi_n|}{1 + (\lambda \beta_n)^2} \sum_{\substack{n=1 \\ n \notin S}}^{\infty} \frac{|\psi_n| |\overline{\eta}_n|}{1 + (\lambda \beta_n)^2}.$$

$$(6.28)$$

For an arbitrary $\beta_0 > 0$, consider the sums in (6.28) for $\beta_n < \beta_0$ and $\beta_n \geq \beta_0$:

$$|\Delta_1| \leq \frac{\lambda k_1}{2} \left(\sum_{\substack{n=1 \\ n \notin S \\ \beta_n < \beta_0}}^{\infty} \frac{\varphi_n^2 \beta_n^2}{\left(1 + (\lambda \beta_n)^2 \right)^2} + \sum_{\substack{n=1 \\ n \notin S \\ \beta_n \geq \beta_0}}^{\infty} \frac{\varphi_n^2 \beta_n^2}{\left(1 + (\lambda \beta_n)^2 \right)^2} \right)^{\frac{1}{2}} \left(\sum_{\substack{n=1 \\ n \notin S}}^{\infty} |\overline{\xi}_n|^2 \right)^{\frac{1}{2}}$$

$$
+ \frac{k_1}{2} \left(\sum_{\substack{n=1 \\ n \notin S}}^{\infty} \frac{|\varphi_n|^2}{\left(1+(\lambda\beta_n)^2\right)^2} \right)^{\frac{1}{2}} \left(\sum_{\substack{n=1 \\ n \notin S}}^{\infty} |\overline{\eta}_n|^2 \right)^{\frac{1}{2}} + \frac{\lambda^2 k_1 k_2}{4} \sum_{\substack{n=1 \\ n \notin S}}^{\infty} \frac{\psi_n^2}{1+(\lambda\beta_n)^2}
$$

$$
\times \left(\sum_{\substack{n=1 \\ n \notin S \\ \beta_n < \beta_0}}^{\infty} \frac{\varphi_n^2 \beta_n^2}{\left(1+(\lambda\beta_n)^2\right)^2} + \sum_{\substack{n=1 \\ n \notin S \\ \beta_n \geq \beta_0}}^{\infty} \frac{\varphi_n^2 \beta_n^2}{\left(1+(\lambda\beta_n)^2\right)^2} \right)^{\frac{1}{2}} \left(\sum_{\substack{n=1 \\ n \notin S}}^{\infty} |\overline{\xi}_n|^2 \right)^{\frac{1}{2}}
$$

$$
+ \frac{\lambda k_1 k_2}{4} \sum_{\substack{n=1 \\ n \notin S}}^{\infty} \frac{\psi_n^2}{1+(\lambda\beta_n)^2} \left(\sum_{\substack{n=1 \\ n \notin S}}^{\infty} \frac{\varphi_n^2}{\left(1+(\lambda\beta_n)^2\right)^2} \right)^{\frac{1}{2}} \left(\sum_{\substack{n=1 \\ n \notin S}}^{\infty} |\overline{\eta}_n|^2 \right)^{\frac{1}{2}}
$$

$$
+ \frac{\lambda^2 k_1 k_2}{4} \left(\sum_{\substack{n=1 \\ n \notin S}}^{\infty} \frac{\psi_n^2}{\left(1+(\lambda\beta_n)^2\right)^2} \right)^{\frac{1}{2}} \left(\sum_{\substack{n=1 \\ n \notin S}}^{\infty} \varphi_n^2 \right)^{\frac{1}{2}}
$$

$$
\times \left(\sum_{\substack{n=1 \\ n \notin S \\ \beta_n < \beta_0}}^{\infty} \frac{\psi_n^2 \beta_n^2}{\left(1+(\lambda\beta_n)^2\right)^2} + \sum_{\substack{n=1 \\ n \notin S \\ \beta_n \geq \beta_0}}^{\infty} \frac{\psi_n^2 \beta_n^2}{\left(1+(\lambda\beta_n)^2\right)^2} \right)^{\frac{1}{2}} \left(\sum_{\substack{n=1 \\ n \notin S}}^{\infty} |\overline{\xi}_n|^2 \right)^{\frac{1}{2}}
$$

$$
+ \frac{\lambda k_1 k_2}{4} \left(\sum_{\substack{n=1 \\ n \notin S}}^{\infty} \frac{\psi_n^2}{\left(1+(\lambda\beta_n)^2\right)^2} \right) \left(\sum_{\substack{n=1 \\ n \notin S}}^{\infty} \varphi_n^2 \right)^{\frac{1}{2}} \left(\sum_{\substack{n=1 \\ n \notin S}}^{\infty} |\overline{\eta}_n|^2 \right)^{\frac{1}{2}} . \tag{6.29}
$$

We have used the Hölder inequality in expression (6.29). By taking into account the inequalities

$$
\left(\sum_{\substack{n=1 \\ n \notin S}}^{\infty} |\overline{\xi}_n|^2 \right)^{\frac{1}{2}} = \|\overline{\xi}\|_{\ell^2} \leq \|\overline{x}\|, \qquad \left(\sum_{\substack{n=1 \\ n \notin S}}^{\infty} |\overline{\eta}_n|^2 \right)^{\frac{1}{2}} = \|\overline{\eta}\|_{\ell^2} \leq \|\overline{x}\|,
$$

$$
\left(\sum_{\substack{n=1 \\ n \notin S \\ \beta_n < \beta_0}}^{\infty} \frac{\varphi_n^2 \beta_n^2}{\left(1+(\lambda\beta_n)^2\right)^2} + \sum_{\substack{n=1 \\ n \notin S \\ \beta_n \geq \beta_0}}^{\infty} \frac{\varphi_n^2 \beta_n^2}{\left(1+(\lambda\beta_n)^2\right)^2} \right)^{\frac{1}{2}} \leq \|\varphi\| \left(\beta_0 + \frac{1}{\lambda^4 \beta_0^2} \right)^{\frac{1}{2}},
$$

expression (6.29) takes the following form

$$|\Delta_1| \le \|\bar{x}\|\|\varphi\| \left(\left(\frac{\lambda k_1}{2} + \frac{\lambda^2 k_1 k_2}{2} \|\psi\|^2 \right) \left(\beta_0 + \frac{1}{\lambda^4 \beta_0^2} \right)^{\frac{1}{2}} + \frac{k_1}{2} + \frac{\lambda k_1 k_2}{2} \|\psi\|^2 \right)$$

$$= M_1(\lambda)\|\bar{x}\|.$$

Similarly,

$$|\Delta_2| \le \|\bar{x}\|\|\psi\| \left(\left(\frac{\lambda k_2}{2} + \frac{\lambda^2 k_1 k_2}{2} \|\varphi\|^2 \right) \left(\beta_0 + \frac{1}{\lambda^4 \beta_0^2} \right)^{\frac{1}{2}} + \frac{k_2}{2} + \frac{\lambda k_1 k_2}{2} \|\varphi\|^2 \right)$$

$$= M_2(\lambda)\|\bar{x}\|.$$

To estimate the determinant Δ from the below, we use the representation

$$\Delta = 1 + \lambda r(\lambda), \tag{6.30}$$

where

$$r(\lambda) = -\frac{k_2}{2} \sum_{\substack{n=1 \\ n \notin S}}^{\infty} \frac{\psi_n^2}{1 + (\lambda \beta_n)^2} - \frac{k_1}{2} \sum_{\substack{n=1 \\ n \notin S}}^{\infty} \frac{\varphi_n^2}{1 + (\lambda \beta_n)^2} + \frac{\lambda k_1 k_2}{4} \sum_{\substack{n=1 \\ n \notin S}}^{\infty} \frac{\varphi_n^2}{1 + (\beta_n)^2}$$

$$\times \sum_{\substack{n=1 \\ n \notin S}}^{\infty} \frac{\psi_n^2}{1 + (\lambda \beta_n)^2} - \frac{\lambda k_1 k_2}{4} \left(\sum_{\substack{n=1 \\ n \notin S}}^{\infty} \frac{\varphi_n \psi_n}{1 + (\lambda \beta_n)^2} \right)^2 .$$

Due to representation (6.30), there exist a $\lambda_0 > 0$ such that

$$\Delta > 0 \quad \text{for each } \lambda \in (0, \lambda_0].$$

Thus, for any $\lambda \in (0, \lambda_0]$, there are numbers $M_4(\lambda) > 0$ and $M_5(\lambda) > 0$ such that

$$|v_1(\bar{x})| \le \frac{M_1(\lambda)}{\Delta} \|\bar{x}\| = M_4(\lambda)\|\bar{x}\|, \quad |v_2(\bar{x})| \le \frac{M_2(\lambda)}{\Delta} \|\bar{x}\| = M_5(\lambda)\|\bar{x}\|$$

in formula (6.27) for all $\bar{x} \in \ell^2$. Formulas (6.24) and (6.27) define $x = (\lambda F + I)^{-1}\bar{x}$ for all $\bar{x} \in \ell^2$ and $\lambda \in (0, \lambda_0]$. Hence,

$$\|(\lambda F + I)^{-1}\bar{x}\|^2 \le 2 \sum_{\substack{n=1 \\ n \notin S}}^{\infty} \frac{1}{1 + (\lambda \beta_n)^2} \sum_{\substack{n=1 \\ n \notin S}}^{\infty} \left(\bar{\xi}_n^2 + \left(\bar{\eta}_n - \lambda \varphi_n v_1(\bar{x}) - \lambda \psi_n v_2(\bar{x}) \right)^2 \right)$$

$$\leq 2\|\bar{x}\|^2 \sum_{\substack{n=1 \\ n\notin S}}^{\infty} \frac{1}{1+(\lambda\beta_n)^2}$$

$$\times \left(1 + \lambda^2 M_4^2 \|\varphi\|^2 + \lambda^2 M_5^2 \|\psi\|^2 + 2\lambda^2 M_4 M_5^2 \|\varphi\|\|\psi\| \right.$$

$$\left. - 2\lambda M_4 \|\varphi\| - 2\lambda M_5 \|\psi\| \right). \tag{6.31}$$

Let us show that the series $\sum_{\substack{n=1 \\ n\notin S}}^{\infty} \frac{1}{1+(\lambda\beta_n)^2}$ is convergent in (6.31). For this purpose we use representations (6.16) and (6.17) to estimate the sum with two indices:

$$\sum_{\substack{n=1 \\ n\notin S}}^{\infty} \frac{1}{1+(\lambda\beta_n)^2} \leq \sum_{k,j=1}^{\infty} \frac{1}{1+\lambda^2\lambda_{kj}} \leq \frac{(l_1 l_2)^4}{\lambda^2\alpha^2} \sum_{k,j=1}^{\infty} \frac{1}{\left((\pi k l_2)^2 + (\pi j l_1)^2\right)^2}.$$

Then we use the integral convergence test:

$$\frac{(l_1 l_2)^4}{\lambda^2\alpha^2} \sum_{k=1}^{\infty} \sum_{j=1}^{\infty} \frac{1}{\left((\pi k l_2)^2 + (\pi j l_1)^2\right)^2} \leq \frac{(l_1 l_2)^4}{\lambda^2\alpha^2} \sum_{k=1}^{\infty} \int_0^{\infty} \frac{dj}{\left((\pi k l_2)^2 + (\pi j l_1)^2\right)^2}$$

$$= \frac{(l_1 l_2)^4}{\lambda^2\alpha^2} \sum_{k=1}^{\infty} \left(\frac{1}{4\pi^3 l_2 k^3 l_1^3} \right)$$

$$\leq \frac{(l_1 l_2)^4}{\lambda^2\alpha^2} \left(\frac{1}{4\pi^3 l_2 l_1^3} + \int_1^{\infty} \left(\frac{1}{4\pi^3 l_2 k^3 l_1^3} \right) \right)$$

$$= \frac{3 l_1 l_2^3}{8\pi^3 \lambda^2\alpha^2}. \tag{6.32}$$

Formulas (6.31) and (6.32) imply that

$$\|(\lambda F + I)^{-1}\bar{x}\|^2 \leq M_6(\lambda)\|\bar{x}\|^2.$$

Therefore, we may define the bounded linear operator

$$(\lambda F + I)^{-1} : \quad \ell^2 \to \ell^2$$

for $\lambda > 0$. To prove its compactness, we consider the bounded linear operator $P_N : \ell^2 \to \ell^2$ that projects an element $x \in \ell^2$ onto the finite dimensional subspace with $\xi_n = \eta_n = 0$ as $n < N$:

$$P_N : \begin{pmatrix} \xi_1 \\ \vdots \\ \eta_{N-1} \\ \xi_N \\ \eta_N \\ \xi_{N+1} \\ \eta_{N+1} \\ \vdots \end{pmatrix} \mapsto \begin{pmatrix} 0 \\ \vdots \\ 0 \\ \xi_N \\ \eta_N \\ \xi_{N+1} \\ \eta_{N+1} \\ \vdots \end{pmatrix}.$$

Then we introduce the following bounded linear operator in ℓ^2:

$$U_N = (I - P_N)(\lambda F + I)^{-1}.$$

The operator U_N is compact as its image is finite dimensional. Note that the operator $(\lambda F + I)^{-1}$ is the limit of compact operators with respect to the norm:

$$\lim_{N \to \infty} \|(\lambda F + I)^{-1} - U_N\| = \lim_{N \to \infty} \|P_N(\lambda F + I)^{-1}\| = 0. \qquad (6.33)$$

Then

$$\|P_N(\lambda F + I)^{-1} \bar{x}\| \leq \sum_{\substack{n=N \\ n \notin S}}^{\infty} \frac{2}{1 + (\lambda \beta_n)^2} \sum_{\substack{n=N \\ n \notin S}}^{\infty} \left(\bar{\xi}_n^2 + \left(\bar{\eta}_n - \lambda \varphi_n v_1(\bar{x}) - \lambda \psi_n v_2(\bar{x}) \right)^2 \right)$$

$$\leq 2\|\bar{x}\|^2 \sum_{\substack{n=N \\ n \notin S}}^{\infty} \frac{1}{1 + (\lambda \beta_n)^2}$$

$$\times \left(1 + \lambda^2 M_4^2 \|\varphi\|^2 + \lambda^2 M_5^2 \|\psi\|^2 + 2\lambda^2 M_4 M_5 \|\varphi\| \|\psi\| \right.$$
$$\left. - 2\lambda M_4 \|\varphi\| - 2\lambda M_5 \|\psi\| \right). \qquad (6.34)$$

Formula (6.32) yields

$$\sum_{\substack{n=N \\ n \notin S}}^{\infty} \frac{1}{1 + (\lambda \beta_n)^2} \to 0 \quad \text{as} \quad N \to \infty.$$

Thus, the estimate (6.34) implies property (6.33). Therefore, the operator

$$(\lambda F + I)^{-1} : \ell^2 \to \ell^2$$

is compact as the limit of finite dimensional operators. It follows from the compactness of the linear operator $(\lambda F + I)^{-1}$ that each trajectory of linear differential equation (6.22) is precompact in ℓ^2 due to Theorem 1.6.

Let us now verify condition (iv) of Theorem 2.4. Suppose that $x(t)$, $t \geq 0$ is a solution of system (6.18) with the control $u = v(x(t))$ of form (6.21), and let $\mu(x(\tau)) = 0$ for some $\tau \geq 0$. We denote $\tilde{x}^0 = x(\tau) \in \ell^2$ and define $\tilde{x}(t) = (\tilde{\xi}_1(t), \tilde{\eta}_1(t), \tilde{\xi}_2(t), \tilde{\eta}_2(t), \ldots)^T$ as the solution of the Cauchy problem:

$$\begin{cases} \dot{\tilde{\xi}}_n(t) = \beta_n \tilde{\eta}_n(t), \\ \dot{\tilde{\eta}}_n(t) = -\beta_n \tilde{\xi}_n(t), \end{cases} \qquad \tilde{x}(0) = \tilde{x}^0. \tag{6.35}$$

The property $\mu(\tilde{x}^0) = 0$ means that $\tilde{\xi}_n(0) = \tilde{\eta}_n(0) = 0$ for all $n \in \mathbb{N} \backslash S$. Then

$$\tilde{\xi}_n(t) = \tilde{\eta}_n(t) = 0, \quad \forall n \in \mathbb{N} \backslash S, \quad \forall t \geq 0. \tag{6.36}$$

This implies that $\mu(\tilde{x}(t)) \equiv 0$. Let us show that $\tilde{x}(t)$ is a solution of system (6.18) with feedback control (6.21). By substituting relation (6.36) into the feedback control (6.21), we conclude that $u = v(\tilde{x}(t)) \equiv 0$. Thus, $\tilde{x}(t)$ is a solution of the closed-loop system (6.18) and (6.21). The uniqueness of solutions of the Cauchy problem implies that

$$x(t) = \tilde{x}(t + \tau), \qquad \forall \, t \geq 0.$$

Hence, from the identity (6.36) it follows that $\mu(x(t)) \equiv 0$. Therefore, condition (iv) holds.

It is remain to verify condition (v) of Theorem 2.4. For thus purpose, we will show that each trajectory $\{x(t)\}_{t \geq 0}$ of the closed-loop system (6.18) with control (6.21) possesses the property $\mu(x(t)) \equiv 0$ on the set

$$M = \{x \in \ell^2 \mid \dot{V}(x) = 0\}.$$

Indeed, let $x(t) \in M$ for all $t \geq 0$, i.e.

$$\dot{V}(x(t)) = -k_1 \left(\sum_{\substack{n=1 \\ n \notin S}}^{\infty} \varphi_n \eta_n(t) \right)^2 - k_2 \left(\sum_{\substack{n=1 \\ n \notin S}}^{\infty} \psi_n \eta_n(t) \right)^2 \equiv 0.$$

As $k_1 > 0$ and $k_2 > 0$, we conclude that

$$v_1(x(t)) = v_2(x(t)) = 0. \tag{6.37}$$

The substitution of the control $u = v(x(t))$ of form (6.37) into (6.18) leads to the following system

$$\begin{cases} \dot{\xi}_n(t) = \beta_n \eta_n(t), \\ \dot{\eta}_n(t) = -\beta_n \xi_n(t), \end{cases} \qquad n \in \mathbb{N}. \tag{6.38}$$

The general solution of system (6.38) is

$$\begin{cases} \xi_n(t) = \xi_n(0) \cos(\beta_n t) + \eta_n(0) \sin(\beta_n t), \\ \eta_n(t) = -\xi_n(0) \sin(\beta_n t) + \eta_n(0) \cos(\beta_n t), \end{cases} \quad n \in \mathbb{N}. \tag{6.39}$$

By substituting (6.39) into (6.37), we get

$$\begin{cases} \displaystyle\sum_{\substack{n=1 \\ n \notin S}}^{\infty} (\varphi_n \eta_n(0) \cos(\beta_n t) - \varphi_n \xi_n(0) \sin(\beta_n t)) \equiv 0, \\ \displaystyle\sum_{\substack{n=1 \\ n \notin S}}^{\infty} (\psi_n \eta_n(0) \cos(\beta_n t) - \psi_n \xi_n(0) \sin(\beta_n t)) \equiv 0, \end{cases} \quad \forall t \geq 0. \tag{6.40}$$

Let us note that the identities (6.40) hold only for

$$\xi_n(0) = \eta_n(0) = 0, \quad \forall n \in \mathbb{N} \setminus S, \tag{6.41}$$

provided that the functions

$$\{\cos(\beta_n t), \ \sin(\beta_n t) \mid n \in \mathbb{N} \setminus S\} \tag{6.42}$$

are linearly independent on $t \in [0, \infty)$. The relations (6.41) together with formulas (6.39) lead to the property $\mu(x(t)) \equiv 0$, which proves condition (v) of Theorem 2.4.

To prove property (6.42), we will show that the functions

$$\{\cos(\beta_n t), \ \sin(\beta_n t) \mid n \in \mathbb{N}\} \tag{6.43}$$

are linearly independent on $t \in [0, \infty)$. We use Theorem 1.2.17 from [17] to prove this fact: if

$$\varlimsup_{a \to \infty} \varlimsup_{z \to \infty} \frac{m[a, a+z)}{z} < \frac{\tau}{2\pi}, \tag{6.44}$$

then system (6.43) is minimal in $L^2(0; \tau)$. In (6.44), $m[a, b)$ denotes the number of elements in the set $[a, b) \cap K$, where

$$K = \{\beta_n \mid n \in \mathbb{N}\}.$$

For proving (6.44), we take into account formulas (6.16) and (6.17) to note that

$$\begin{aligned} \beta_n = \sqrt{\lambda_{kj}} &= \alpha \pi^2 \left(\frac{k^2}{l_1^2} + \frac{j^2}{l_2^2} \right) \\ &= \frac{\alpha \pi^2}{l_2^2} \left(j^2 + k^2 \chi \right) \text{ for } (k, j) = (k_n, j_n), \ \chi = \frac{l_2^2}{l_1^2} > 0. \end{aligned}$$

Fig. 6.2 Sets Ω, B_+, and B_-

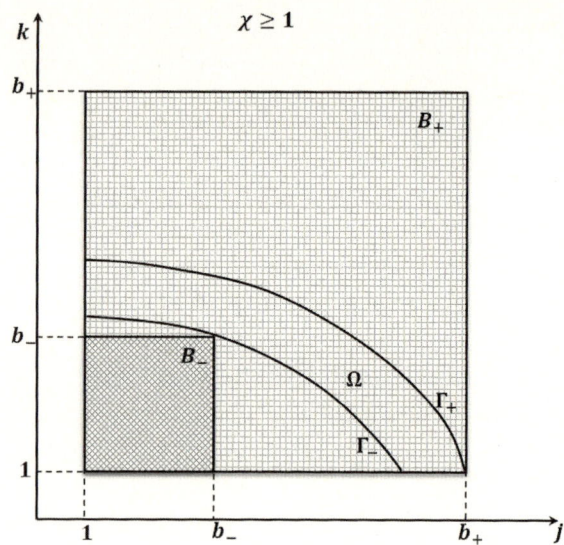

Let $\tilde{\beta}_{kj} = j^2 + k^2\chi$. Consider first the case $\chi \geq 1$.

Let $\Omega \subset \mathbb{R}^2$ be a domain containing the points $(j, k) \in \mathbb{N}^2$ such that $\tilde{\beta}_{kj} \in [a, b)$,

$$\Omega = \{(j, k) \mid a \leq j^2 + k^2\chi < b, \quad k \geq 1, \ j \geq 1\}.$$

We denote components of the boundary of Ω by $\Gamma_+ : j^2 + k^2\chi = b$ and $\Gamma_- : j^2 + k^2\chi = a$. Thus, the estimate of $m[a, b)$ is reduced to computing the number of elements in the set $\Omega \cap \mathbb{N}^2$. Consider a square $B_+ \supset \Omega$ and a square B_- such that $B_- \cap \Omega$ contains at most one point (see Fig. 6.2).

Let us find the number of integer-valued points in squares B_+ and B_-. For this purpose we compute the coordinates of vertices for B_+ and B_-:

$$b_+ : b_+^2 + \chi 1^2 = b, \qquad b_- : b_-^2 + \chi b_-^2 = a. \tag{6.45}$$

From these formulas we deduce that $b_+ = \sqrt{b - \chi}$ and $b_- = \sqrt{\dfrac{a}{1 + \chi}}$. Then the number of integer-valued points in Ω can be estimated as follows:

$$m[a; b) \leq |B_+ \cap \mathbb{N}^2| - |B_- \cap \mathbb{N}^2| + 1 = b_+^2 - b_-^2 + 1 = b - \chi - \dfrac{a}{1 + \chi} + 1.$$

Let us denote $b = a + z$, then

$$\varlimsup_{a \to \infty} \varlimsup_{z \to \infty} \dfrac{m[a, a + z)}{z} \leq \varlimsup_{a \to \infty} \varlimsup_{z \to \infty} \dfrac{a\left(1 - \dfrac{1}{1 + \chi}\right) + z - \chi + 1}{z} = 1. \tag{6.46}$$

Similarly, for the case $\chi < 1$

$$\lim_{a \to \infty} \lim_{z \to \infty} \frac{-m[a, a+z]}{z} \leq \lim_{a \to \infty} \lim_{z \to \infty} \frac{a\left(\dfrac{1}{\chi} - \dfrac{1}{1+\chi}\right) + \dfrac{z-1}{\chi} + 1}{z} = \frac{1}{\chi}. \quad (6.47)$$

From estimates (6.46) and (6.47) we conclude that functions (6.43) are linearly independent on $[0, \tau]$ (as condition (6.44) holds) if $\tau > 2\pi \max\left\{1, \dfrac{1}{\chi}\right\}$.

Thus, the solution $x = 0$ of the closed-loop system (6.18) with the control $u = v(x)$ is asymptotically stable with respect to the functional μ by Theorem 2.4.

6.3 Finite Dimensional Controllability

Let us fix an integer number $N \geq 1$ and consider a finite dimensional subsystem of (6.18) with N blocks:

$$\begin{pmatrix} \dot{\xi}_i \\ \dot{\eta}_i \end{pmatrix} = A_i \begin{pmatrix} \xi_i \\ \eta_i \end{pmatrix} + B_i \begin{pmatrix} u_1 \\ u_2 \end{pmatrix}, \quad i = 1, 2, ..., N. \quad (6.48)$$

Controllability conditions for system (6.48) are given by the following theorem.

Theorem 6.2 *System (6.48) is controllable if*

$$\beta_i \neq 0 \quad \text{for all } i = 1, 2, ..., N, \quad (6.49)$$

and the following condition hold:

$$\begin{vmatrix} \varphi_1 & \varphi_2 & \cdots & \varphi_N \\ \beta_1^2 \varphi_1 & \beta_2^2 \varphi_2 & \cdots & \beta_N^2 \varphi_N \\ \vdots & \vdots & \ddots & \vdots \\ \beta_1^{2(p-1)} \varphi_1 & \beta_2^{2(p-1)} \varphi_2 & \cdots & \beta_N^{2(p-1)} \varphi_N \\ \psi_1 & \psi_2 & \cdots & \psi_N \\ \beta_1^2 \psi_1 & \beta_2^2 \psi_2 & \cdots & \beta_N^2 \psi_N \\ \vdots & \vdots & \ddots & \vdots \\ \beta_1^{2(q-1)} \psi_1 & \beta_2^{2(q-1)} \psi_2 & \cdots & \beta_N^{2(q-1)} \psi_N \end{vmatrix} \neq 0 \quad (6.50)$$

for some non-negative integers p and q such that $p + q = N$. (In case $p = 0$ the matrix in (6.50) does not contain the rows with φ_i, and in case $q = 0$ there are no rows with ψ_i.)

Proof We use the Kalman criterion to prove this theorem. Let us verify that

$$\mathrm{rank}(B, AB, ..., A^{2N-1}B) = 2N.$$

Matrices A and B have the form

$$A = \begin{pmatrix} A_1 & 0 & \cdots & 0 \\ 0 & A_2 & \cdots & 0 \\ \vdots & \vdots & \ddots & \vdots \\ 0 & 0 & \cdots & A_N \end{pmatrix} \in \mathrm{mat}(2N \times 2N), \quad B = \begin{pmatrix} B_1 \\ B_2 \\ \vdots \\ B_N \end{pmatrix} \in \mathrm{mat}(2N \times 2),$$

where A_i and B_i are given by formulas (6.19) Let us compute the matrix $K = (B, AB, ..., A^{2N-1}B)$. The products $AB, A^2B, ..., A^{2N-1}B$ are presented as follows

$$AB = \begin{pmatrix} \beta_1\varphi_1 & \beta_1\psi_1 \\ 0 & 0 \\ \beta_2\varphi_2 & \beta_2\psi_2 \\ 0 & 0 \\ \vdots & \vdots \\ \vdots & \vdots \\ \beta_N\varphi_N & \beta_N\psi_N \end{pmatrix}, \quad A^2B = \begin{pmatrix} 0 & 0 \\ -\beta_1^2\varphi_1 & -\beta_1^2\psi_1 \\ 0 & 0 \\ -\beta_2^2\varphi_2 & -\beta_2^2\psi_2 \\ \vdots & \vdots \\ \vdots & \vdots \\ 0 & 0 \\ -\beta_N^2\varphi_N & -\beta_N^2\psi_N \end{pmatrix},$$

$$A^{2N-1}B = \begin{pmatrix} (-1)^{N-1}\beta_1^{2N-1}\varphi_1 & (-1)^{N-1}\beta_1^{2N-1}\psi_1 \\ 0 & 0 \\ (-1)^{N-1}\beta_2^{2N-1}\varphi_2 & (-1)^{N-1}\beta_2^{2N-1}\psi_2 \\ 0 & 0 \\ \vdots & \vdots \\ \vdots & \vdots \\ (-1)^{N-1}\beta_N^{2N-1}\varphi_N & (-1)^{N-1}\beta_N^{2N-1}\psi_N \\ 0 & 0 \end{pmatrix}.$$

By substituting these values into the matrix $K = (B, AB, ..., A^{2N-1}B)$, we get:

$$K = \begin{pmatrix} 0 & 0 & \cdots\cdots & (-1)^{N-1}\beta_1^{2N-1}\varphi_1 & (-1)^{N-1}\beta_1^{2N-1}\psi_1 \\ \varphi_1 & \psi_1 & \cdots\cdots & 0 & 0 \\ 0 & 0 & \cdots\cdots & (-1)^{N-1}\beta_2^{2N-1}\varphi_2 & (-1)^{N-1}\beta_2^{2N-1}\psi_2 \\ \varphi_2 & \psi_2 & \cdots\cdots & 0 & 0 \\ \vdots & \vdots & \ddots\cdots & \vdots & \vdots \\ \vdots & \vdots & \cdots\ddots & \vdots & \vdots \\ 0 & 0 & \cdots\cdots & (-1)^{N-1}\beta_N^{2N-1}\varphi_N & (-1)^{N-1}\beta_N^{2N-1}\psi_N \\ \varphi_N & \psi_N & \cdots\cdots & 0 & 0 \end{pmatrix}.$$

Let us consider two cases.

(1) For simplicity, consider first the case $p = 0$. In this case we choose the rows and columns of K with coefficients ψ_{kj} and write the $2N \times 2N$-matrix thus obtained:

$$K^* = \begin{pmatrix} 0 & \beta_1\psi_1 & 0 & \cdots & (-1)^{N-1}\beta_1^{2N-1}\psi_1 \\ \psi_1 & 0 & -\beta_1^2\psi_1 & \cdots & 0 \\ 0 & \beta_2\psi_2 & 0 & \cdots & (-1)^{N-1}\beta_2^{2N-1}\psi_2 \\ \vdots & \vdots & \vdots & \ddots & \vdots \\ \psi_N & 0 & -\beta_N^2\psi_N & \cdots & 0 \end{pmatrix}.$$

To find the rank of K^*, we compute its determinant by transforming K^* to a block form by the permutation of rows and columns:

$$\begin{vmatrix} 0 & \beta_1\psi_1 & 0 & \cdots & (-1)^{N-1}\beta_1^{2N-1}\psi_1 \\ \psi_1 & 0 & -\beta_1^2\psi_1 & \cdots & 0 \\ 0 & \beta_2\psi_2 & 0 & \cdots & (-1)^{N-1}\beta_2^{2N-1}\psi_2 \\ \vdots & \vdots & \vdots & \ddots & \vdots \\ \psi_N & 0 & -\beta_N^2\psi_N & \cdots & 0 \end{vmatrix}$$

$$= \prod_{p=1}^{N}(\psi_p^2\beta_p) \begin{vmatrix} 0 & 0 & \cdots & 0 & 1 & -\beta_1^2 & \cdots & (-1)^{N-1}\beta_1^{2N-1} \\ \vdots & \vdots & \ddots & \vdots & 1 & -\beta_2^2 & \cdots & (-1)^{N-1}\beta_2^{2N-1} \\ \vdots & \vdots & \ddots & \vdots & \vdots & \vdots & \ddots & \vdots \\ 0 & 0 & \cdots & 0 & 1 & -\beta_N^2 & \cdots & (-1)^{N-1}\beta_N^{2N-1} \\ 1 & -\beta_1^2 & \cdots & (-1)^{N-1}\beta_1^{2N-1} & 0 & 0 & \cdots & 0 \\ 1 & -\beta_2^2 & \cdots & (-1)^{N-1}\beta_2^{2N-1} & \vdots & \vdots & \ddots & \vdots \\ \vdots & \vdots & \ddots & \vdots & \vdots & \vdots & \ddots & \vdots \\ 1 & -\beta_N^2 & \cdots & (-1)^{N-1}\beta_N^{2N-1} & 0 & 0 & \cdots & 0 \end{vmatrix}$$

$$= \prod_{p=1}^{N}\left(\psi_p^2\beta_p\right) \prod_{1\le l<n\le N}\left(\beta_l^2 - \beta_n^2\right)^2 \ne 0.$$

We see that the blocks of the above determinant correspond to the determinant of form (6.50) with $p = 0$. We have also used here the fact that

$$\begin{vmatrix} 1 & 1 & \cdots & 1 \\ \beta_1^2 & \beta_2^2 & \cdots & \beta_N^2 \\ \vdots & \vdots & \ddots & \vdots \\ \beta_1^{2N-1} & \beta_2^{2N-1} & \cdots & \beta_N^{2N-1} \end{vmatrix} = \prod_{1\le l<n\le N}\left(\beta_l^2 - \beta_n^2\right)$$

is the Vandermonde determinant.

Thus, $\text{rank}(K^*) = 2N$ under conditions (6.49) and (6.50) with $p = 0$, so that system (6.48) is controllable by the Kalman criterion.

(2) Now we consider the general case $p + q = N$ and apply similar transformations. We choose the rows and columns of K with coefficients of the form

$$\beta_n^{2i}\,\varphi_n,\ \ \beta_n^{2l}\psi_n \ \ \text{for } i = 0, ..., p - 1,\ l = 0, ..., q - 1,$$

and consider the $2N \times 2N$-matrix K^* thus obtained. Proceeding similarly to case (1), we conclude that $\det K^* \neq 0$ provided that conditions (6.49) and (6.50) hold. So, system (6.48) is controllable by the Kalman criterion.

Remark 6.1 Theorem 6.2 is valid for any system of form (6.48) without assuming any special form for the coefficients β_n, φ_n, and ψ_n of matrices A_n and B_n in formulas (6.19). However, if these coefficients are given by relations (6.16) with taking into account notations (6.17), then system (6.48) is not controllable if N is large enough. Namely, if a one-to-one correspondence $\mathbb{N} \ni n \mapsto (k_n, j_n) \in \mathbb{N}^2$ in (6.17) is given, then there is an $n \in \mathbb{N}$ such that both corresponding j_n and k_n are even. Then by formulas (6.16) we conclude that $\varphi_n = b_{k_n,j_n}^1 = 0$ and $\psi_n = b_{k_n,j_n}^2 = 0$. It means that condition (6.50) of Theorem 6.2 is not satisfied for any system (6.48) with $N \geq n$. Indeed, in this case the n-th column of the matrix in (6.50) is zero, and condition (6.50) fails. We see that system (6.48) is not controllable in this case as it admits the integral $\xi_n^2(t) + \eta_n^2(t) = \text{const}$.

6.4 Spectral Controllability

To overcome the obstruction faced in Remark 6.1, we will study a restriction of system (6.15) to the subspace spanned by the coordinates that satisfy controllability conditions (6.49) and (6.50). For this purpose we consider the set

$$L = \{(k, j) \in \mathbb{N}^2 \mid j \text{ is odd or } k \text{ is odd}\}.$$

Then we introduce a one-to-one correspondence $n \mapsto (k_n, j_n)$ between $\mathbb{N}$ and L. In Sects. 6.4 and 6.5, we use the following notations with index $n \in \mathbb{N}$ defined through the double index $(k_n, j_n) \in L$:

$$\xi_n := P_{kj},\ \ \eta_n := Q_{kj},\ \ \beta_n := \sqrt{\lambda_{kj}} > 0,\ \ \varphi_n := b_{kj}^1,$$
$$\psi_n := b_{kj}^2 \ \text{for } (k, j) = (k_n, j_n) \in L. \tag{6.51}$$

With these notations we write a subsystem of system (6.15) as follows:

$$\dot{x} = Ax + Bu,\ \ x \in \ell^2,\ \ u \in \mathbb{R}^2, \tag{6.52}$$

where

$$x = \begin{pmatrix} \xi_1 \\ \eta_1 \\ \xi_2 \\ \eta_2 \\ \vdots \end{pmatrix}, \quad A = \begin{pmatrix} A_1 & 0 & \cdots \\ 0 & A_2 & \cdots \\ \vdots & \vdots & \ddots \end{pmatrix}, \quad B = \begin{pmatrix} B_1 \\ B_2 \\ \vdots \end{pmatrix}, \quad u = \begin{pmatrix} u_1 \\ u_2 \end{pmatrix}.$$

We assume that A_n and B_n are given by (6.19), and the parameters $\lambda_{kj}, b^1_{kj}, b^2_{kj}$ are defined by formulas (6.16).

Note that the construction of system (6.52) differs from system (6.18) considered in Sect. 6.2. Namely, the index n in representations (6.17) and (6.51) parameterizes different sets $\mathbb{N}^2$ and L, respectively.

We recall that an infinite dimensional system is *spectrally controllable* if all its finite dimensional subsystems are controllable (cf. [15]). For system (6.52), we may formally consider any its subsystem (6.52) for $N = 1, 2, \ldots$ Thus, system (6.52) is called *spectrally controllable* if, for any number $N \geq 1$, its finite dimensional subsystem of form (6.48) is controllable.

A straightforward corollary of Theorem 6.2 is the following result about the spectral controllability of system (6.52).

Theorem 6.3 *Assume that the matrices A_n and B_n of system (6.52) are given by (6.19), their parameters are defined by formulas (6.51), (6.16), and*

$$\chi = \frac{l_2^2}{l_1^2}$$

is an irrational number. Then system (6.52) is spectrally controllable provided that $2a_1 \neq l_1$ *or* $2a_2 \neq l_2$.

Proof From representations (6.51) and (6.16) we conclude that

$$\beta_n = \sqrt{\lambda_{kj}} = \alpha\pi^2 \left(\frac{k^2}{l_1^2} + \frac{j^2}{l_2^2} \right) - \frac{\alpha\pi^2}{l_2^2} \left(j^2 + k^2\chi \right) > 0, \tag{6.53}$$

and

$$\varphi_n = b^1_{k_n, j_n} \neq 0 \text{ (if } 2a_2 \neq l_2) \text{ or } \psi_n b^2_{k_n, j_n} \neq 0 \text{ (if } 2a_1 \neq l_1) \tag{6.54}$$

for all $n \in \mathbb{N}$, $(k_n, j_n) \in L$ due to the construction of the set L.

For an arbitrary $N \geq 1$, condition (6.49) of Theorem 6.2 is satisfied due to representation (6.53). If $2a_2 \neq l_2$ then $\varphi_n \neq 0$ for all n because of (6.54). Then we take $p = N$ and $q = 0$ in condition (6.50) of Theorem 6.2:

$$
\begin{vmatrix}
\varphi_1 & \varphi_2 & \cdots & \varphi_N \\
\beta_1^2 \varphi_1 & \beta_2^2 \varphi_2 & \cdots & \beta_N^2 \varphi_N \\
\vdots & \vdots & \ddots & \vdots \\
\beta_1^{2(N-1)} \varphi_1 & \beta_2^{2(N-1)} \varphi_2 & \cdots & \beta_N^{2(N-1)} \varphi_N
\end{vmatrix}
= \varphi_1 \varphi_2 \cdots \varphi_N \prod_{1 \le l < n \le N} \left(\beta_l^2 - \beta_n^2 \right).
$$

Therefore, to prove that this determinant does not vanish, it is enough to show that

$$
\beta_l \ne \beta_n \quad \text{for all } l \ne n. \tag{6.55}
$$

We see that property (6.55) holds due to representation (6.53) and the assumption that the number χ is irrational. Hence, condition (6.50) of Theorem 6.2 holds if $2a_2 \ne l_2$. If $2a_1 \ne l_1$, we use an analogous construction to verify condition (6.50) with $p = 0$ and $q = N$ as all $\psi_n \ne 0$ in this case due to (6.54).

Thus, system (6.52) is spectrally controllable by Theorem 6.2.

6.5 Optimal Control Problem

In this section, we fix a number $N \ge 1$ and consider the following optimal control problem for a subsystem of (6.52).

Problem Statement. For given $x^0 \in \mathbb{R}^{2N}$, $x^1 \in \mathbb{R}^{2N}$, and $\tau > 0$, find a control $u \in L^2 \left((0, \tau); \mathbb{R}^2 \right)$ that minimizes the functional

$$
J = \int_0^\tau \left(u_1^2(t) + u_2^2(t) \right) dt \to \min \tag{6.56}
$$

on the solutions $x(t)$, $t \in [0, \tau]$ of system

$$
\dot{x}(t) = \bar{A} x(t) + \bar{B} u(t), \quad x(t) \in \mathbb{R}^{2N}, \ u(t) \in \mathbb{R}^2, \tag{6.57}
$$

subject to the boundary conditions

$$
x(0) = x^0, \ x(\tau) = x^1. \tag{6.58}
$$

The components of system (6.57) and boundary conditions (6.58) are presented as follows:

$$
x = \begin{pmatrix} x_1 \\ x_2 \\ \vdots \\ x_N \end{pmatrix}, \quad x^i = \begin{pmatrix} x_1^i \\ x_2^i \\ \vdots \\ x_N^i \end{pmatrix}, \quad x_n = \begin{pmatrix} \xi_n \\ \eta_n \end{pmatrix}, \quad x_n^i = \begin{pmatrix} \xi_n^i \\ \eta_n^i \end{pmatrix},
$$

$$
i = 0, 1, \quad n = 1, 2, \ldots, N,
$$

$$\bar{A} = \begin{pmatrix} A_1 & 0 & \cdots & 0 \\ 0 & A_2 & \cdots & 0 \\ \vdots & \vdots & \ddots & \vdots \\ 0 & 0 & \cdots & A_N \end{pmatrix}, \quad \bar{B} = \begin{pmatrix} B_1 \\ B_2 \\ \vdots \\ B_N \end{pmatrix},$$

where A_n and B_n are defined in (6.19) with coefficients given by formulas (6.16) and (6.51).

The main result of this section is as follows.

Theorem 6.4 *Assume that the conditions of Theorem 6.2 are satisfied for system (6.57). Then, for arbitrary $\tau > 0$ and boundary conditions (6.58), there is a unique optimal control $u = \hat{u}(t)$ for the problem (6.56)–(6.58):*

$$\begin{cases} \hat{u}_1(t) = \sum_{i=1}^{N} \varphi_i \left(q_i^0 \cos(\beta_i t) - p_i^0 \sin(\beta_i t) \right), \\ \hat{u}_2(s) = \sum_{i=1}^{N} \psi_i \left(q_i^0 \cos(\beta_i t) - p_i^0 \sin(\beta_i t) \right). \end{cases} \tag{6.59}$$

Here the parameters p_i^0, q_i^0 are defined by the following system of linear algebraic equations:

$$\left(\tfrac{\tau}{2} I + F \right) \begin{pmatrix} q_1^0 \\ p_1^0 \\ \vdots \\ q_{k_n j_N}^0 \\ p_{k_n j_N}^0 \end{pmatrix} = P, \tag{6.60}$$

where

$$F = \begin{pmatrix} F_{11} & F_{12} & \cdots & F_{1N} \\ F_{21} & F_{22} & \cdots & F_{2N} \\ \vdots & \vdots & \ddots & \vdots \\ F_{N1} & F_{N2} & \cdots & F_{NN} \end{pmatrix}, \quad F_{ll} = \begin{pmatrix} \dfrac{\sin(2\beta_l \tau)}{4\beta_l} & -\dfrac{\sin^2(\beta_l \tau)}{2\beta_l} \\ -\dfrac{\sin^2(\beta_l \tau)}{2\beta_l} & -\dfrac{\sin(2\beta_l \tau)}{4\beta_l} \end{pmatrix},$$

$$F_{li} = \frac{\varphi_l \varphi_i + \psi_l \psi_i}{(\varphi_l^2 + \psi_l^2)(\beta_l^2 - \beta_i^2)} \begin{pmatrix} F_{li}^{11} & F_{li}^{12} \\ F_{li}^{21} & F_{li}^{22} \end{pmatrix},$$

$$F_{li}^{11} = \beta_l \sin(\beta_l \tau) \cos(\beta_i \tau) - \beta_i \cos(\beta_l \tau) \sin(\beta_i \tau),$$

$$F_{li}^{12} = -(\beta_l \sin(\beta_l \tau) \sin(\beta_i \tau) + \beta_i \cos(\beta_l \tau) \cos(\beta_i \tau) - \beta_i),$$

$$F_{li}^{21} = \beta_l \cos(\beta_l \tau) \cos(\beta_i \tau) + \beta_i \sin(\beta_l \tau) \sin(\beta_i \tau) - \beta_l,$$

$$F_{li}^{22} = -(\beta_l \cos(\beta_l \tau) \sin(\beta_i \tau) - \beta_i \cos(\beta_i \tau) \sin(\beta_l \tau)),$$

$$P = \begin{pmatrix} \dfrac{\xi_1^1 \sin(\beta_1 \tau) + \eta_1^1 \cos(\beta_1 \tau) - \eta_1^0}{\varphi_1^2 + \psi_1^2} \\ \dfrac{\xi_1^1 \cos(\beta_1 \tau) - \eta_1^1 \sin(\beta_1 \tau) - \xi_1^0}{\varphi_1^2 + \psi_1^2} \\ \vdots \\ \dfrac{\xi_{k_n j_N}^1 \sin(\beta_{k_n j_N} \tau) + \eta_{k_n j_N}^1 \cos -\eta_{k_n j_N}^0}{\varphi_{k_n j_N}^2 + \psi_{k_n j_N}^2} \\ \dfrac{\xi_{k_n j_N}^1 \cos(\beta_{k_n j_N} \tau) - \eta_{k_n j_N}^1 \sin(\beta_{k_n j_N} \tau) - \xi_{k_n j_N}^0}{\varphi_{k_n j_N}^2 + \psi_{k_n j_N}^2} \end{pmatrix}.$$

Proof In order to solve the optimal control problem (6.56)–(6.58), we apply the Pontryagin maximum principle with the following Hamiltonian (cf. [18]):

$$H(\zeta_0, \zeta, x, u) = \zeta_0(u_1^2 + u_2^2) + \sum_{i=1}^{N} (p_i \beta_i \eta_i + q_i(-\beta_i \xi_i + \varphi_i u_1 + \psi_i u_2))$$
$$= \zeta_0(Qu, u) + \zeta(\bar{A}x + \bar{B}u),$$

where

$$Q = \begin{pmatrix} 1 & 0 \\ 0 & 1 \end{pmatrix}, \quad \zeta = (p_1, q_1, ..., p_N, q_N).$$

Without loss of generality we assume that $\zeta_0 = -1/2$. The optimal control $u = \hat{u}(t)$ can be founded from the maximization of the Hamiltonian. For this purpose we find critical points of H from the equation $\dfrac{\partial H}{\partial u} = 0$ with respect to u, i.e. $2\zeta_0 Q\hat{u} + \zeta\bar{B} = 0$. This yields

$$\hat{u}(t) = -\frac{1}{2\zeta_0}Q^{-1}\zeta(t)\bar{B} = \begin{pmatrix} \sum\limits_{i=1}^{N} \varphi_i q_i(t) \\ \sum\limits_{i=1}^{N} \psi_i q_i(t) \end{pmatrix}. \tag{6.61}$$

We consider expression (6.61) for $\zeta(t)$ that satisfies the Hamiltonian system

$$\dot{x} = \frac{\partial H}{\partial \zeta}\bigg|_{u=\hat{u}} = \bar{A}x + \bar{B}\hat{u}(t), \quad \dot{\zeta} = -\frac{\partial H}{\partial x}\bigg|_{u=\hat{u}} = -\zeta\bar{A}. \tag{6.62}$$

We write the component of system (6.62) with respect to $\zeta(t)$ as

$$\begin{cases} \dot{p}_i(t) = \beta_i q_i(t), \\ \dot{q}_i(t) = -\beta_i p_i(t). \end{cases} \quad i = \overline{1, N} \tag{6.63}$$

The general solution of the above system is

$$\begin{cases} p_i(t) = p_i^0 \cos(\beta_i(t)) + q_i^0 \sin(\beta_i(t)), \\ q_i(t) = -p_i^0 \sin(\beta_i(t)) + q_i^0 \cos(\beta_i(t)). \end{cases} \tag{6.64}$$

By substituting solution (6.64) into expression (6.61), we obtain the control of form (6.59).

Let us find constants p_i^0 and q_i^0 from the boundary conditions (6.58). For this purpose we represent the solution $x(t)$ of system (6.57) with the control $\hat{u}(t)$ as follows:

$$x_l(t) = e^{tA_l} x_l^0 + \int_0^\tau e^{(t-s)A_l} B_l \hat{u}(s) ds, \tag{6.65}$$

where $e^{tA_l} x_l^0 = \begin{pmatrix} \cos(\beta_l t) & \sin(\beta_l t) \\ -\sin(\beta_l t) & \cos(\beta_l t) \end{pmatrix} \begin{pmatrix} \xi_l^0 \\ \eta_l^0 \end{pmatrix}$. Expression (6.65) can be represented as

$$\xi_l(t) = \xi_l^0 \cos(\beta_l t) + \eta_l^0 \sin(\beta_l t)$$

$$+ \int_0^t \left(\sum_{i=1}^N \varphi_l \varphi_i \left(q_i^0 \cos(\beta_i s) - p_i^0 \sin(\beta_i s) \right) \right.$$

$$\left. + \psi_l \sum_{i=1}^N \psi_i \left(q_i^0 \cos(\beta_i s) - p_i^0 \sin(\beta_i s) \right) \right) \sin(\beta_l(t-s)) ds,$$

$$\eta_l(t) = -\xi_l^0 \sin(\beta_l t) + \eta_l^0 \cos(\beta_l t)$$

$$+ \int_0^t \left(\sum_{i=1}^N \varphi_l \varphi_i \left(q_i^0 \cos(\beta_i s) - p_i^0 \sin(\beta_i s) \right) \right.$$

$$\left. + \psi_l \sum_{i=1}^n \psi_i \left(q_i^0 \cos(\beta_i s) - p_i^0 \sin(\beta_i s) \right) \right) \cos(\beta_l(t-s)) ds.$$

$$\tag{6.66}$$

From the boundary conditions, we take

$$\xi_l(\tau) = \xi_l^1, \quad \eta_l(\tau) = \eta_l^1. \tag{6.67}$$

After computing integrals in (6.66), relation (6.67) takes the form

$$\sum_{i \neq l, i=1}^{N} q_i^0 (\varphi_l \varphi_i + \psi_l \psi_i) \beta_l \frac{\cos(\beta_i \tau) - \cos(\beta_l \tau)}{\beta_l^2 - \beta_i^2} + q_l^0 (\varphi_l^2 + \psi_l^2) \frac{\tau \sin(\beta_l \tau)}{2}$$

$$- \sum_{i \neq l, i=1}^{N} p_i^0 (\varphi_l \varphi_i + \psi_l \psi_i) \frac{\beta_l \sin(\beta_i \tau) - \beta_i \sin(\beta_l \tau)}{\beta_l^2 - \beta_i^2} - p_l^0 (\varphi_l^2 + \psi_l^2)$$

$$\times \frac{\sin(\beta_l \tau) - \beta_l \tau \cos(\beta_l \tau)}{2 \beta_l} = \xi_l^1 - \xi_l^0 \cos(\beta_l \tau) - \eta_l^0 \sin(\beta_l \tau),$$

$$\sum_{i \neq l, i=1}^{N} q_i^0 (\varphi_l \varphi_i + \psi_l \psi_i) \beta_l \frac{\sin(\beta_l \tau) - \beta_i \sin(\beta_i \tau)}{\beta_l^2 - \beta_i^2}$$

$$+ q_l^0 (\varphi_l^2 + \psi_l^2) \frac{\sin(\beta_l \tau) + \beta_l \tau \cos(\beta_l \tau)}{2 \beta_l} - \sum_{i \neq l, i=1}^{N} p_i^0 (\varphi_l \varphi_i + \psi_l \psi_i)$$

$$\times \frac{\cos(\beta_i \tau) - \cos(\beta_l \tau)}{\beta_l^2 - \beta_i^2} \beta_i - p_l^0 (\varphi_l^2 + \psi_l^2) \frac{\tau \sin(\beta_l \tau)}{2}$$

$$= \eta_l^1 + \xi_l^0 \sin(\beta_l \tau) - \eta_l^0 \cos(\beta_l \tau). \tag{6.68}$$

Let us transform system (6.68) as follows: we multiply the first equation by $\sin(\beta_l \tau)$ and add the result to the second equation multiplied by $\cos(\beta_l \tau)$. Then we multiply the first equation by $\cos(\beta_l \tau)$ and add to it the second equation multiplied by $-\sin(\beta_l \tau)$. Then system (6.68) takes the form:

$$q_l^0 \frac{\tau}{2} + q_l^0 \frac{\sin(2\beta_l \tau)}{4\beta_l} - p_l^0 \frac{\sin^2(\beta_l \tau)}{2\beta_l} + \sum_{i \neq l, i=1}^{N} q_i^0 \frac{(\varphi_l \varphi_i + \psi_l \psi_i)}{\varphi_l^2 + \psi_l^2}$$

$$\times \left(\frac{\beta_l \sin(\beta_l \tau) \cos(\beta_i \tau) - \beta_i \cos(\beta_l \tau) \sin(\beta_i \tau)}{\beta_l^2 - \beta_i^2} \right)$$

$$- \sum_{i \neq l, i=1}^{N} p_i^0 \frac{(\varphi_l \varphi_i + \psi_l)}{\psi_i} \varphi_l^2$$

$$+ \psi_l^2 \left(\frac{\beta_l \sin(\beta_l \tau) \sin(\beta_i \tau) + \beta_i \cos(\beta_l \tau) \cos(\beta_i \tau) - \beta_i}{\beta_l^2 - \beta_i^2} \right)$$

$$= \frac{\xi_l^1 \sin(\beta_l \tau) + \eta_l^1 \cos(\beta_l \tau) - \eta_l^0}{\varphi_l^2 + \psi_l^2},$$

$$p_l^0 \frac{\tau}{2} - p_l^0 \frac{\sin(2\beta_l \tau)}{4\beta_l} - q_l^0 \frac{\sin^2(\beta_l \tau)}{2\beta_l} + \sum_{i \neq l, i=1}^{N} q_i^0 \frac{(\varphi_l \varphi_i + \psi_l \psi_i)}{\varphi_l^2 + \psi_l^2}$$

$$\times \left(\frac{\beta_l \cos(\beta_l \tau) \cos(\beta_i \tau) + \beta_i \sin(\beta_l \tau) \sin(\beta_i \tau) - \beta_l}{\beta_l^2 - \beta_i^2} \right)$$

$$- \sum_{i \neq l, i=1}^{N} p_i^0 \frac{(\varphi_l \varphi_i + \psi_l \psi_i)}{\varphi_l^2 + \psi_l^2}$$

$$\times \left(\frac{\beta_l \cos(\beta_l \tau) \sin(\beta_i \tau) - \beta_i \cos(\beta_i \tau) \sin(\beta_l \tau)}{\beta_l^2 - \beta_i^2} \right)$$

$$= \frac{\xi_l^1 \cos(\beta_l \tau) - \eta_l^1 \sin(\beta_l \tau) - \xi_l^0}{\varphi_l^2 + \psi_l^2}.$$

The above system of linear algebraic equations with respect to p_i^0, q_i^0 takes the form (6.60) in matrix notation. Note that system (6.57) is controllable under our assumptions.

Now the assertion of Theorem 6.4 follows from the Pontryagin maximum principle and existence results for the linear regulator problem (see, e.g., [18]).

6.6 Numerical Simulations

As an example, we consider a rotating rigid body with the Kirchhoff plate for the following mechanical parameters:

$$l_1 = 1 \, \text{m}, \ l_2 = \pi \, \text{m}, \ a_1 = -a_2 = 1 \, \text{m}, \ h = 0.01 \, \text{m},$$
$$\alpha^2 = 1 \, \text{N} \, \text{m}^3/\text{kg}, \ I_\rho = 0 \, \text{kg}.$$

Then the first nine eigenfrequencies

$$\sqrt{\lambda_{kj}} = \alpha \left(\left(\frac{\pi k}{l_1} \right)^2 + \left(\frac{\pi j}{l_2} \right)^2 \right), \quad 1 \leq k, j \leq 3,$$

for system (6.15) are as follows:

$$\sqrt{\lambda_{11}} \approx 10.869 \, \text{s}^{-1}; \ \sqrt{\lambda_{12}} \approx 13.869 \, \text{s}^{-1}, \ \sqrt{\lambda_{13}} \approx 18.869 \, \text{s}^{-1},$$

$$\sqrt{\lambda_{21}} \approx 40.475 \, \text{s}^{-1}, \ \sqrt{\lambda_{22}} \approx 43.474 \, \text{s}^{-1}, \ \sqrt{\lambda_{23}} \approx 48.474 \, \text{s}^{-1},$$

$$\sqrt{\lambda_{31}} \approx 89.809 \, \text{s}^{-1}, \ \sqrt{\lambda_{32}} \approx 92.808 \, \text{s}^{-1}, \ \sqrt{\lambda_{33}} \approx 97.807 \, \text{s}^{-1}.$$

We consider the corresponding subsystem of (6.15) with nine modes of plate vibrations:

$$\frac{d}{dt}\begin{pmatrix} P_{kj}(t) \\ Q_{kj}(t) \end{pmatrix} = \begin{pmatrix} 0 & \sqrt{\lambda_{kj}} \\ -\sqrt{\lambda_{kj}} & 0 \end{pmatrix}\begin{pmatrix} P_{kj}(t) \\ Q_{kj}(t) \end{pmatrix} + \begin{pmatrix} 0 & 0 \\ b_{kj}^1 & b_{kj}^2 \end{pmatrix}\begin{pmatrix} u_1(t) \\ u_2(t) \end{pmatrix}, \quad 1 \le k, j \le 3,$$

$$\tag{6.69}$$

where the coefficients b_{kj}^1 and b_{kj}^2 are defined by formulas (6.16).

To illustrate the efficiency of the optimal control obtained in Theorem 6.4, consider a subsystem of system (6.69) with three degrees o freedom. For this purpose we introduce a parametrization

$$n \mapsto (k_n, j_n), \quad n = 1, 2, 3, \tag{6.70}$$

and changes of variables of form (6.17):

$$\xi_n := P_{kj}, \ \eta_n := Q_{kj}, \ \beta_n := \sqrt{\lambda_{kj}} > 0, \ \varphi_n := b_{kj}^1,$$

$$\psi_n := b_{kj}^2 \quad \text{for } (k, j) = (k_n, j_n). \tag{6.71}$$

We construct the map $n \mapsto (k_n, j_n)$ in (6.70) to select the three minimal eigenfrequencies $\beta_1 \le \beta_2 \le \beta_3$} from $\sqrt{\lambda_{kj}}$. This requirement results in the following map of form (6.70):

$$(k_n, j_n) = \begin{cases} (1, 1), & \text{for } n = 1, \\ (1, 2), & \text{for } n = 2, \\ (1, 3), & \text{for } n = 3. \end{cases}$$

Then $\beta_1 \approx 10.869\,\text{s}^{-1}$, $\beta_2 \approx 13.869\,\text{s}^{-1}$, and $\beta_3 \approx 18.869\,\text{s}^{-1}$.

Consider the cost functional

$$J = \int_0^1 \{u_1^2(t) + u_2^2(t)\}dt \to \min \tag{6.72}$$

for the subsystem of system (6.69) corresponding to notations (6.71):

$$\begin{pmatrix} \dot{\xi}_n \\ \dot{\eta}_n \end{pmatrix} = \begin{pmatrix} 0 & \omega_n \\ -\omega_n & 0 \end{pmatrix}\begin{pmatrix} \xi_n \\ \eta_n \end{pmatrix} + \begin{pmatrix} 0 & 0 \\ \phi_n & \psi_n \end{pmatrix}\begin{pmatrix} u_1 \\ u_2 \end{pmatrix}, \quad n = 1, 2, 3, \tag{6.73}$$

and boundary conditions

$$\xi_1(0) = 1, \ \xi_1(1) = \xi_2(0) = \xi_2(1) = \xi_3(0) = \xi_3(1) = 0, \ \eta_n(0) = \eta_n(1) = 0. \tag{6.74}$$

The solution of linear system (6.60) corresponding to the optimal control problem (6.72)–(6.74) is as follows:

$$p_1^0 \approx -1.291, \quad p_2^0 \approx -0.848, \quad p_3^0 \approx 1.242,$$

$$q_1^0 \approx 0.071, \quad q_2^0 \approx -3.141, \quad q_3^0 \approx 0.750.$$

By using these constants p_n^0 and q_n^0, we apply the control $u = \hat{u}(t)$ of form (6.59) to system (6.69) with the following initial conditions:

$$P_{11}(0) = 1, \quad Q_{11}(0) = 0, \quad P_{kj}(0) = Q_{kj}(0) = 0, \quad \text{for } (k, j) \in \{1, 2, 3\}^2 \setminus \{(1, 1)\}.$$

The time plot of the norm $\|x(t)\|$ of the corresponding solution of system (6.69) is shown in Fig. 6.3, where

$$\|x(t)\| = \sum_{k,j=1}^{3} \left(P_{kj}^2(t) + Q_{kj}^2(t) \right)^{1/2}.$$

As we see in Fig. 6.3, the norm of $x(t)$ is close to zero at $t = \tau = 1$ for system (6.69) with nine elastic modes. This observation illustrates that the L^2-minimal control corresponding to the reduced system with three low-frequency modes (6.73) can be used for (approximate) solving the steering problem of the multi-dimensional system (6.69). We will justify the applicability of optimal controls, corresponding to finite dimensional subsystems, for the approximate controllability problem in the next section.

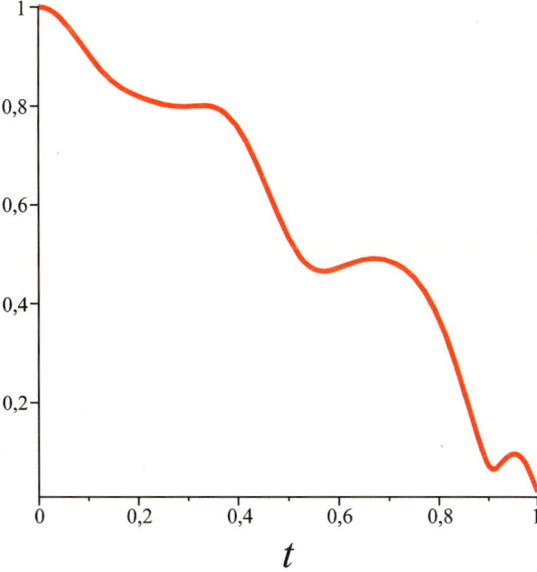

Fig. 6.3 The graph of $\|x(t)\|, t \in [0, 1]$

6.7 Reachable Sets of the Single-Input Model

Let us consider system (6.15) with the scalar control $u_1 = u \in \mathbb{R}$ by assuming that $u_2 \equiv 0$. From formulas (6.16) we see that all the coefficients b^1_{kj} vanish for even k, so we will control the subsystem of (6.15) corresponding to odd indices k. For this purpose we introduce the set of indices

$$\hat{L} = \{(k, j) \in \mathbb{N}^2 \mid k \text{ is odd}\}$$

and assume that there is a one-to-one correspondence

$$n \mapsto (k_n, j_n) \tag{6.75}$$

between the sets $\mathbb{N}$ and $\hat{L}$. Then we introduce the following complex variables

$$q_n = P_{k_n j_n} + i Q_{k_n j_n}, \quad q_{-n} = \bar{q}_n = P_{k_n j_n} - i Q_{k_n j_n},$$

and coefficients

$$\omega_n = \sqrt{\lambda_{k_n j_n}} = \alpha \left(\left(\frac{\pi k_n}{l_1} \right)^2 + \left(\frac{\pi j_n}{l_2} \right)^2 \right), \tag{6.76}$$

$$B_n = b^1_{k_n j_n} = \begin{cases} \dfrac{2\sqrt{l_1 l_2^3}}{\pi^2 k_n j_n}, & \text{for even } j_n, \\ \dfrac{2\sqrt{l_1 l_2}(2a_2 - l_2)}{\pi^2 k_n j_n}, & \text{for odd } j_n, \end{cases} \tag{6.77}$$

$$B_{-n} = -B_n, \quad n = 1, 2, \ldots,$$

where P_{kj} and Q_{kj} are state variables of system (6.15). Let us assume that $2a_2 \neq l_2$, and that the map (6.75) is chosen in such a way that

$$0 < \omega_1 \leq \omega_2 \leq \ldots \leq \omega_n \leq \ldots.$$

By differentiating each $q_n(t)$ along the trajectories of system (6.15) for $n = \pm 1$, $\pm 2, \ldots$, we obtain the following control system:

$$\dot{q} = Aq + Bu, \quad q \in \ell^2, \quad u \in \mathbb{R}^1, \tag{6.78}$$

where

$$
q = \begin{pmatrix} q_{-1} \\ q_1 \\ q_{-2} \\ q_2 \\ \vdots \end{pmatrix} \in \ell^2, \quad A = i \begin{pmatrix} \omega_1 & 0 & 0 & 0 & \cdots \\ 0 & -\omega_1 & 0 & 0 & \cdots \\ 0 & 0 & \omega_2 & 0 & \cdots \\ 0 & 0 & 0 & -\omega_2 & \cdots \\ \vdots & \vdots & \vdots & \vdots & \ddots \end{pmatrix}, \quad B = i \begin{pmatrix} B_{-1} \\ B_1 \\ B_{-2} \\ B_2 \\ \vdots \end{pmatrix}.
$$

System (6.78) is considered in the Hilbert space ℓ^2 with the norm

$$
\|q\|_{\ell^2} = \left(\sum_{n=1}^{\infty} \left(|q_n|^2 + |q_{-n}|^2 \right) \right)^{1/2}.
$$

Operator $A : D(A) \to \ell^2$ is the infinitesimal generator of a C_0-semigroup of linear operators $\{e^{tA}\}_{t \geq 0}$ on ℓ^2 due to the Hille–Yosida theorem (Theorem 1.2). Thus, for any $q^0 \in \ell^2$, $\tau > 0$, and $u \in L^2(0, \tau)$, there is a unique mild solution $q(t, q^0, u)$ of differential equation (6.78) with $u = u(t)$, $t \in [0, \tau]$, satisfying the initial condition $q|_{t=0} = q^0$. Such a solution can be described by the formula

$$
q(t; q^0, u) = e^{tA} q^0 + \int_0^t e^{(t-s)A} B u(s) ds, \quad 0 \leq t \leq \tau.
$$

Consider the reachable sets for system (6.78):

$$
R_\tau(q^0) = \{ q^1 \in \ell^2 \mid q^1 = q(\tau; q^0, u) \text{ for } u \in L^2(0, \tau) \},
$$
$$
R(q^0) = \bigcup_{\tau \geq 0} R_\tau(q^0).
$$

Let us recall that system (6.78) is called to be *approximately controllable* if $\overline{R(q^0)} = \ell^2$ for all $q^0 \in \ell^2$. A standard method for the study of the approximate controllability of systems of type (6.78) is based on the result due to Levan and Rigby [19]. This approach relies on the analysis of invariant subspaces of the adjoint semigroup $\{e^{tA^*}\}_{t \geq 0}$ in the kernel of B^*. Up to our knowledge, direct applications of that approach, as well as the concept of simultaneous controllability studied in the monograph [20], do not allow estimating the reachable sets $R_\tau(q^0)$ and constructing controls that solve the steering problem for given boundary conditions. In this section, we will follow the approach described in Chap. 4 and use optimal controls corresponding to finite-dimensional problems to estimate the reachable sets $R_\tau(q^0)$ for system (6.78).

In addition to system (6.78), we consider its finite dimensional subsystem corresponding to coordinates q_{-n}, q_n, $n = \overline{1, N}$, for a fixed integer number $N \geq 1$:

$$\dot{\tilde{q}}_N = A_N \tilde{q}_N + B_N u, \tag{6.79}$$

$$A_N = i \begin{pmatrix} \omega_1 & 0 & \cdots & 0 & 0 \\ 0 & -\omega_1 & \cdots & 0 & 0 \\ \vdots & \vdots & \ddots & \vdots & \vdots \\ 0 & 0 & \cdots & \omega_N & 0 \\ 0 & 0 & \cdots & 0 & -\omega_N \end{pmatrix}, \quad \tilde{q}_N = \begin{pmatrix} q_{-1} \\ q_1 \\ \vdots \\ q_{-N} \\ q_N \end{pmatrix}, \quad B_N = i \begin{pmatrix} B_{-1} \\ B_1 \\ \vdots \\ B_{-N} \\ B_N \end{pmatrix}.$$

Theorem 6.4 can be reformulated for system (6.79) with complex state variables as follows (see also [21]).

Theorem 6.5 *Let $\omega_j \neq \omega_k$ for all $1 \leq j < k \leq N$. Consider the following optimal control problem:*

$$\dot{\tilde{q}}_N = A_N \tilde{q}_N + B_N u, \quad t \in [0, \tau], \tag{6.80}$$

$$J = \int_0^\tau |u(t)|^2 dt \longrightarrow min, \tag{6.81}$$

$$\tilde{q}_N^0 = \tilde{q}_N(0) = \begin{pmatrix} q_{-1}^0 \\ q_1^0 \\ \vdots \\ q_{-N}^0 \\ q_N^0 \end{pmatrix} \in \mathbb{C}^{2N}, \quad \tilde{q}_N^1 = \tilde{q}_N(\tau) = \begin{pmatrix} q_{-1}^1 \\ q_1^1 \\ \vdots \\ q_{-N}^1 \\ q_N^1 \end{pmatrix} \in \mathbb{C}^{2N}, \tag{6.82}$$

$$q_n^0 = \overline{q_{-n}^0}, \quad q_n^1 = \overline{q_{-n}^1}, \quad n = \overline{1, N}.$$

The optimal control for this problem has the form

$$\hat{u}_N(t) = (B_1 e^{i\omega_1 t}, B_{-1} e^{-i\omega_1 t}, ..., B_N e^{i\omega_N t}, B_{-N} e^{-i\omega_N t})\nu,$$

where

$$\nu = \begin{pmatrix} \nu_{-1} \\ \nu_1 \\ \vdots \\ \nu_{-N} \\ \nu_N \end{pmatrix} = \begin{pmatrix} \frac{1}{B_1} & 0 & \cdots & 0 & 0 \\ 0 & \frac{1}{B_{-1}} & \cdots & 0 & 0 \\ \vdots & \vdots & \ddots & \vdots & \vdots \\ 0 & 0 & \cdots & \frac{1}{B_N} & 0 \\ 0 & 0 & \cdots & 0 & \frac{1}{B_{-N}} \end{pmatrix} K^{-1}$$

$$\times \begin{pmatrix} \frac{1}{B_{-1}} & 0 & \cdots & 0 & 0 \\ 0 & \frac{1}{B_1} & \cdots & 0 & 0 \\ \vdots & \vdots & \ddots & \vdots & \vdots \\ 0 & 0 & \cdots & \frac{1}{B_{-N}} & 0 \\ 0 & 0 & \cdots & 0 & \frac{1}{B_N} \end{pmatrix} (e^{-i\omega_N \tau} \tilde{q}_N^1 - \tilde{q}_N^0), \tag{6.83}$$

$$K = (K_{jk})_{j,k=1}^N, \qquad K_{jj} = \begin{pmatrix} \tau & \dfrac{i(e^{-2i\omega_j \tau} - 1)}{2\omega_j} \\ \dfrac{i(1 - e^{2i\omega_j \tau})}{2\omega_j} & \tau \end{pmatrix},$$

$$K_{jk} = i \begin{pmatrix} \dfrac{1 - e^{i(\omega_k - \omega_j)\tau}}{\omega_k - \omega_j} & \dfrac{e^{-i(\omega_k + \omega_j)\tau} - 1}{\omega_k + \omega_j} \\ \dfrac{1 - e^{i(\omega_k + \omega_j)\tau}}{\omega_k + \omega_j} & \dfrac{1 - e^{i(\omega_j - \omega_k)\tau}}{\omega_j - \omega_k} \end{pmatrix}, \qquad j \neq k.$$

To study reachable sets for system (6.78), let us recall that a complex number χ is called an *algebraic number* if it is a root of a non-zero polynomial in one variable with integer coefficients [22]. A number n^* is called the order of an algebraic number χ, if χ is a root of some n^*-th order polynomial with integer coefficients and there are no other non-zero polynomials with integer coefficients of order less than n^* having the root χ.

Let us formulate the main result of this section on the estimate of states $q^1 \in \ell^2$ of system (6.78), which are approximately reachable from a point $q^0 = 0 \in \ell^2$.

Theorem 6.6 *Assume that the following conditions are satisfied for system* (6.78):

(1) $B_n \neq 0, n = \pm 1, \pm 2, \pm 3, \ldots$;

(2) $\chi = \left(\dfrac{l_2}{l_1}\right)^2$ *is an algebraic number of order* $n^* \geq 2$;

(3) *vector* $q^1 = (q_{-1}^1, q_1^1, q_{-2}^1, q_2^1, \ldots)^T \in \ell^2$ *admits the following estimate*

$$q_{-n}^1 = \overline{q_n^1}, \ |q_n^1| = O\left(\frac{1}{n^\gamma}\right), \ \gamma > \frac{3}{2}n^* + 1 \quad \text{for } n = 1, 2, \ldots.$$

Then, for any $\varepsilon > 0$, *there are numbers* $\tau = \tau(\varepsilon) > 0$ *and* $N(\varepsilon) \geq 1$ *such that*

$$\|q(\tau; 0, \hat{u}_N) - q^1\|_{\ell^2} < \varepsilon, \tag{6.84}$$

where $\hat{u}_N(t)$ *is the optimal control for problem* (6.80)–(6.82):

$$\hat{u}_N(t) = \sum_{n=1}^{N}(B_n e^{i\omega_n t} \nu_{-n} + B_{-n} e^{-i\omega_n t} \nu_n), \quad t \in [0, \tau].$$

Proof Under conditions of this theorem, χ is an irrational number. Therefore, representation (6.76) implies the property $\omega_j \neq \omega_k$ for all $j \neq k$. According to the idea of Theorem 4.1, we estimate the value of (6.84) by using controls of form $u = \hat{u}_N(t)$ from Theorem 6.5.

Let us introduce operators $Q_N : \ell^2 \to \ell^2$ and $P_N : \ell^2 \to \ell^2$:

$$P_N : q = \begin{pmatrix} q_{-1} \\ q_1 \\ \vdots \\ q_{-N} \\ q_N \\ q_{-N-1} \\ q_{N+1} \\ \vdots \end{pmatrix} \longmapsto \begin{pmatrix} q_{-1} \\ q_1 \\ \vdots \\ q_{-N} \\ q_N \\ 0 \\ 0 \\ \vdots \end{pmatrix}, \quad Q_N = I - P_N.$$

Then

$$\|q(\tau; 0, \hat{u}_N) - q^1\|_{\ell^2} = \|Q_N q(\tau; 0, \hat{u}_N) - Q_N q^1\|$$

$$\leq \|Q_N q^1\| + \left\| \int_0^{\tau} Q_N e^{(\tau-s)A} B \hat{u}_N(s) ds \right\|.$$

Since the operator e^{tA} commutes with Q_N, the Cauchy–Schwartz inequality implies that

$$\|q(\tau; 0, \hat{u}_N) - q^1\|_{\ell^2} \leq \|Q_N q^1\| + \sup_{t \in [0,\tau]} \|e^{tA}\| \cdot \|Q_N B\| \cdot \int_0^{\tau} |\hat{u}_N(s)| ds$$

$$\leq \|Q_N q^1\| + \sqrt{\tau} \|Q_N B\| \cdot \sup_{t \in [0,\tau]} \|e^{tA}\| \cdot \|\hat{u}_N\|_{L^2(0,\tau)}.$$

Note that the norm of operators $e^{tA} : \ell^2 \to \ell^2$ is equal to 1 for all $t \geq 0$. Thus, it is suffices to show that, for any $\varepsilon > 0$ and vector $q^1 \in \ell^2$ satisfying condition (3), there are $\tau > 0$ and N such that

$$\|Q_N q^1\| < \frac{\varepsilon}{2}, \tag{6.85}$$

$$r_N = \tau \|Q_N B\|^2 \|\hat{u}_N\|^2_{L^2(0;\tau)} < \frac{\varepsilon^2}{4}. \tag{6.86}$$

Inequality (6.85) follows from the fact that

$$\lim_{N\to\infty}\|Q_N q^1\|^2 = \lim_{N\to\infty}\sum_{n=N+1}^{\infty}\left(|q_{-n}^1|^2 + |q_n^1|^2\right) = 0,$$

for each element $q^1 \in \ell^2$. To prove inequality (6.86), we compute the norm of optimal controls:

$$\|\hat{u}_N(t)\|_{L^2(0,\tau)}^2 = \int_0^\tau |\hat{u}_N(t)|^2 dt$$

$$= \int_0^\tau \hat{u}_N(t)\overline{\hat{u}_N(t)}dt$$

$$= \int_0^\tau \left(\sum_{n=1}^N B_n e^{i\omega_n t}\nu_{-n} + B_{-n}e^{-i\omega_n t}\nu_n\right)$$

$$\times \left(\sum_{n'=1}^N \bar{B}_{n'} e^{i\omega_{n'}t}\bar{\nu}_{-n'} + \bar{B}_{-n'}e^{-i\omega_{n'}t}\bar{\nu}_{n'}\right)dt$$

$$= \sum_{n=1}^N \left(B_n \bar{B}_n \nu_{-n}\bar{\nu}_{-n}\tau + B_n \bar{B}_{-n}\nu_{-n}\bar{\nu}_n\frac{i(1-e^{2i\omega_n\tau})}{2\omega_n}\right.$$

$$+ B_{-n}\bar{B}_n\nu_n\bar{\nu}_{-n}\frac{i(e^{-2i\omega_n\tau}-1)}{2\omega_n} + \left.\sum_{n=1}^N B_{-n}\bar{B}_{-n}\nu_n\bar{\nu}_n\tau\right)$$

$$+ \sum_{n=1}^N\sum_{n'=1}^N \left(B_n\bar{B}_{n'}\nu_{-n}\bar{\nu}_{-n'}\frac{i(1-e^{i(\omega_n-\omega_{n'})\tau})}{\omega_n-\omega_{n'}}\right.$$

$$+ B_n\bar{B}_{-n'}\nu_{-n}\bar{\nu}_{n'}\frac{i(1-e^{i(\omega_n+\omega_{n'})\tau})}{\omega_n+\omega_{n'}}$$

$$+ B_{-n}\bar{B}_{n'}\nu_n\bar{\nu}_{-n'}\frac{i(e^{-i(\omega_n+\omega_{n'})\tau}-1)}{\omega_n+\omega_{n'}}$$

$$+ \left.B_{-n}\bar{B}_{-n'}\nu_n\bar{\nu}_{n'}\frac{i(e^{i(\omega_{n'}-\omega_n)\tau}-1)}{\omega_n-\omega_{n'}}\right).$$

Let us estimate $\|\hat{u}_N(t)\|_{L^2(0,\tau)}^2$ by using the triangle and the Hölder inequalities:

$$\|\hat{u}_N(t)\|^2_{L^2(0,\tau)} \leq 4 \sum_{\substack{n,n'=1,\\ n\neq n'}}^{N} |B_n B_{n'} \nu_n \nu_{n'}| \left(\frac{1}{|\omega_n - \omega_{n'}|} + \frac{1}{|\omega_n + \omega_{n'}|} \right)$$

$$+ 2 \sum_{n=1}^{N} |B_n|^2 |\nu_n|^2 \left(\tau + \frac{1}{|\omega_n|} \right)$$

$$\leq 2 \left(\tau + \max_{n \leq N} \frac{1}{|\omega_n|} \right) \sum_{n=1}^{N} |B_n|^2 |\nu_n|^2$$

$$+ 4 \left(\max_{n,n' \leq N} \frac{1}{|\omega_n - \omega_{n'}|} + \max_{n,n' \leq N} \frac{1}{|\omega_n + \omega_{n'}|} \right) \sum_{\substack{n,n'=1,\\ n\neq n'}}^{N} |B_n B_{n'} \nu_n \nu_{n'}|$$

$$\leq 2 \max \left\{ \left(\tau + \frac{1}{\omega_1} \right), \left(\frac{2}{\min_{n,n' \leq N} |\omega_n - \omega_{n'}|} + \frac{2}{\omega_1 + \omega_2} \right) \right\}$$

$$\times \sum_{\substack{n,n'=1,\\ n\neq n'}}^{N} |B_n B_{n'} \nu_n \nu_{n'}|$$

$$\leq \frac{4}{\min_{n,n' \leq N} |\omega_n - \omega_{n'}|} \sum_{\substack{n,n'=1,\\ n\neq n'}}^{N} |B_n B_{n'} \nu_n \nu_{n'}|. \tag{6.87}$$

By using transformations

$$\tilde{\nu}_n = \nu_n B_{-n}, \quad \tilde{\nu}_{-n} = \nu_{-n} B_n, \quad \frac{2}{\min_{n,n' \leq N} |\omega_n - \omega_{n'}|} = H(N),$$

estimate (6.87) takes the form

$$\|\hat{u}_N(t)\|^2_{L^2(0,\tau)} \leq 2H(N) \sum_{\substack{n,n'=1,\\ n\neq n'}}^{N} |\tilde{\nu}_n \tilde{\nu}_{n'}|$$

$$= 2H(N) \sum_{n'=1}^{N} \left(\sum_{n=1}^{N} |\tilde{\nu}_n| |\tilde{\nu}_{n'}| \right)$$

$$\leq 2H(N) \sum_{n'=1}^{N} \left(\sqrt{\sum_{n=1}^{N} |\tilde{\nu}_{n'}|^2} \sqrt{\sum_{n=1}^{N} |\tilde{\nu}_n|^2} \right)$$

$$= H(N)\sqrt{2N}\|\tilde{\nu}\|_2 \sum_{n'=1}^{N} |\tilde{\nu}_{n'}|$$

$$\leq H(N)\sqrt{2N}\|\tilde{\nu}\|_2 \sqrt{\sum_{n'=1}^{N} |\tilde{\nu}_{n'}|^2} \sqrt{\sum_{n'=1}^{N} 1^2}$$

$$= H(N)N\|\tilde{\nu}\|_2^2. \tag{6.88}$$

Here $\|\tilde{\nu}\|_2 = \sum_{n=1}^{N} \left(|\tilde{\nu}_n|^2 + |\tilde{\nu}_{-n}|^2\right)^{\frac{1}{2}}$ denotes the Euclidean norm of the vector $\tilde{\nu}$. According to Theorem 6.5,

$$\tilde{\nu} = K^{-1}y, \tag{6.89}$$

where
$$\tilde{\nu} = \begin{pmatrix} \tilde{\nu}_{-1} \\ \tilde{\nu}_1 \\ \vdots \\ \tilde{\nu}_{-N} \\ \tilde{\nu}_N \end{pmatrix}, \quad y = \begin{pmatrix} \frac{1}{B_{-1}} & 0 & \cdots & 0 & 0 \\ 0 & \frac{1}{B_1} & \cdots & 0 & 0 \\ \vdots & \vdots & \ddots & \vdots & \vdots \\ 0 & 0 & \cdots & \frac{1}{B_{-N}} & 0 \\ 0 & 0 & \cdots & 0 & \frac{1}{B_N} \end{pmatrix} (e^{-i\omega_N \tau}\tilde{q}_N^1 - \tilde{q}_N^0).$$

To estimate the norm of $\tilde{\nu}$, we represent the matrix K as $K = \tau I + C$ and consider C as a linear operator from the space $\mathbb{C}^{2N}$ with the norm $\|\cdot\|_1$ to $\mathbb{C}^{2N}$ with the norm $\|\cdot\|_\infty$,

$$C : \left(\mathbb{C}^{2N}, \|\cdot\|_1\right) \to \left(\mathbb{C}^{2N}, \|\cdot\|_\infty\right),$$

where $\|y\|_1 = \sum_{n=1}^{N} \left(|y_n| + |y_{-n}|\right)$, $\|y\|_\infty = \max_{1 \leq |n| \leq N} |y_n|$. Let us rewrite (6.89) in the form

$$K\tilde{\nu} = (\tau I + C)\tilde{\nu} = y, \quad \tilde{\nu} = \frac{y}{\tau} - \frac{C\tilde{\nu}}{\tau},$$

hence,

$$\|\tilde{\nu}\|_\infty \leq \frac{1}{\tau}\left(\|y\|_\infty + \|C\tilde{\nu}\|_\infty\right) \leq \frac{1}{\tau}\left(\|y\|_\infty + \|C\|\|\tilde{\nu}\|_1\right),$$

$$\|\tilde{\nu}\|_\infty \left(1 - \frac{\|C\|}{\tau}\right) \leq \frac{1}{\tau}\|y\|_\infty,$$

$$\|\tilde{\nu}\|_\infty \leq \frac{\|y\|_\infty}{\tau - \|C\|}, \quad \text{provided that} \quad \|C\| < \tau. \tag{6.90}$$

Let us estimate the norm of the operator C:

$$\|C\| \leq \max_{n,m \leq N} |C_{nm}| \leq \max_{1 \leq n \leq N} \left\{ \frac{1}{\omega_n}, 2 \max_{1 \leq m \leq N} \left\{ \frac{1}{|\omega_m - \omega_n|}, \frac{1}{|\omega_m + \omega_n|} \right\} \right\}. \quad (6.91)$$

From formula (6.76), we take

$$\omega_n = \alpha \left(\left(\frac{\pi j_n}{l_1} \right)^2 + \left(\frac{\pi k_n}{l_2} \right)^2 \right), \quad \omega_m = \alpha \left(\left(\frac{\pi j_m}{l_1} \right)^2 + \left(\frac{\pi k_m}{l_2} \right)^2 \right),$$

and introduce the notation

$$\beta_{jk} = \sqrt{\lambda_{jk}} = \alpha \left(\left(\frac{\pi j}{l_1} \right)^2 + \left(\frac{\pi k}{l_2} \right)^2 \right).$$

Inequalities $1 \leq n \leq N$ and $1 \leq m \leq N$ imply that the components of the map (6.75) satisfy constraints $1 \leq j_m, k_m, j_n, k_n \leq M(N)$ for some integer $M(N)$,

$$M(N) = O(\sqrt{N}), \quad \text{as} \quad N \to \infty. \quad (6.92)$$

Then expression (6.91) can be estimated in the following way:

$$\|C\| \leq \max_{j_n, k_n \leq M} \left\{ \frac{1}{\beta_{j_n k_n}}, \max_{\substack{(j_m, k_m) \neq (j_n, k_n) \\ j_m, k_m \leq M}} \left\{ \frac{2}{|\beta_{j_n k_n} + \beta_{j_m k_m}|}, \frac{2}{|\beta_{j_n k_n} - \beta_{j_m k_m}|} \right\} \right\}$$

$$\leq \max \left\{ \frac{1}{\min_{j_n, k_n \leq M} \beta_{j_n k_n}}, \frac{2}{\min_{\substack{(j_m, k_m) \neq (j_n, k_n) \\ j_n, k_n, j_m, k_m \leq M}} |\beta_{j_n k_n} + \beta_{j_m k_m}|}, \right.$$

$$\left. \frac{2}{\min_{\substack{(j_m, k_m) \neq (j_n, k_n) \\ j_n, k_n, j_m, k_m \leq M}} |\beta_{j_n k_n} - \beta_{j_m k_m}|} \right\}.$$

Let

$$j_n = p, \quad k_n = c, \quad j_m = m, \quad k_m = s, \quad \chi = \left(\frac{l_2}{l_1} \right)^2,$$

then

$$\min_{\substack{(j_m, k_m) \neq (j_n, k_n) \\ j_n, k_n, j_m, k_m \leq M}} |\beta_{j_n k_n} - \beta_{j_m k_m}| = \frac{\alpha \pi^2}{l_2^2} \min_{\substack{(p,c) \neq (m,s) \\ p, c, m, s \leq M}} |\chi p^2 + c^2 - \chi m^2 - s^2|.$$

Since χ is an algebraic number of order $n^* \geq 2$, then Liouville's theorem on diophantine approximation yields (cf. [22]):

$$\frac{\alpha\pi^2}{l_2^2} \min_{\substack{(p,c)\neq(m,s) \\ p,c,m,s\leq M}} |\chi p^2 + c^2 - \chi m^2 - s^2| \geq \min_{\substack{p\neq m \\ p,m\leq M}} \frac{\alpha\pi^2 R}{l_2^2|p^2 - m^2|^{n^*-1}}$$

$$\geq \frac{\alpha\pi^2 R}{l_2^2(M^2(N) - 1)^{n^*-1}},$$

where R is a positive constant[2] depending on χ. Therefore,

$$\|C\| \leq \frac{l_2^2(M^2(N) - 1)^{n^*-1}}{\alpha\pi^2 R}.$$

Then formulas (6.88), (6.90), and (6.92) imply

$$\|\hat{u}_N(t)\|^2_{L^2(0,\tau)} \leq NH(N)\|y\|_2^2 \quad \text{if} \quad \tau > \frac{l_2^2(M^2(N) - 1)^{n^*-1}}{\alpha\pi^2 R} = O(N^{n^*-1}).$$

$$(6.93)$$

Let us estimate the Euclidean norm of y:

$$\|y\|_2^2 = \sum_{n=1}^{N} \left(|y_{-n}|^2 + |y_n|^2\right)$$

$$= \sum_{n=1}^{N} \frac{|e^{-i\omega_n\tau}q_n^1 - q_n^0|^2 + |e^{i\omega_n\tau}q_{-n}^1 - q_{-n}^0|^2}{|B_n|^2}$$

$$\leq 2\sum_{n=1}^{N} \frac{|e^{-i\omega_n\tau}q_n^1|^2 + |e^{i\omega_n\tau}q_{-n}^1|^2 + |q_n^0|^2 + |q_{-n}^0|^2}{|B_n|^2}$$

$$\leq 2\sum_{n=1}^{N} \frac{|q_n^1|^2 + |q_{-n}^1|^2 + |q_n^0|^2 + |q_{-n}^0|^2}{|B_n|^2}.$$

By using this expression in (6.86), we get

$$r_N \leq \frac{4N\tau}{\min_{n,n'\leq N}|\omega_n - \omega_{n'}|} \sum_{n=1}^{N} \frac{|q_n^1|^2 + |q_{-n}^1|^2 + |q_n^0|^2 + |q_{-n}^0|^2}{|B_n|^2}$$

$$\times \sum_{n=N+1}^{\infty} \left(|B_{-n}|^2 + |B_n|^2\right).$$

$$(6.94)$$

---

[2] This constant R can be computed explicitly in terms of the value conjugated with χ.

For the map $n \mapsto (k_n, j_n)$ described in (6.75) and numbers $n, n' \in \mathbb{N}$, we denote $(p, c) = (k_n, j_n)$ and $(k, s) = (k_{n'}, j_{n'})$. According to our notations, $\omega_n = \beta_{pc}$ and $\omega_{n'} = \beta_{ks}$, where

$$\beta_{pc} = \alpha\left(\left(\frac{\pi p}{l_1}\right)^2 + \left(\frac{\pi c}{l_2}\right)^2\right), \quad \beta_{ks} = \alpha\left(\left(\frac{\pi k}{l_1}\right)^2 + \left(\frac{\pi s}{l_2}\right)^2\right),$$

and b_{pc}^1 is given in (6.16). Then formula (6.94) reads as

$$r_N \leq \frac{8N\tau}{\min\limits_{\substack{(p,c)\neq(k,s) \\ p,c,k,s \leq N}} |\beta_{pc} - \beta_{ks}|} \sum_{p,c=1}^{N} \frac{|q_n^1|^2 + |q_{-n}^1|^2 + |q_n^0|^2 + |q_{-n}^0|^2}{|B_n|^2} \sum_{p,c=N+1}^{\infty} |i\varphi_{pc}|^2.$$

$$(6.95)$$

Let

$$q_n^0 = 0, \quad q_{-n}^0 = 0, \quad |q_n^1| = |q_{-n}^1| = O\left(\frac{1}{p^\gamma} + \frac{1}{c^\gamma}\right).$$

By using notations (6.77) for B_n, formula (6.95) gives the following representation

$$r_N = O\left(\frac{N\tau \sum_{p,c=1}^{N} \frac{(c^\gamma + p^\gamma)^2}{(pc)^{2\gamma-2}} \cdot \sum_{p,c=N+1}^{\infty} (pc)^{-2}}{\min\limits_{\substack{(p,c)\neq(k,s) \\ p,c,k,s \leq N}} \left|\frac{\alpha\pi^2}{(l_1 l_2)^2}\left(l_1^2(c^2 - s^2) + l_2^2(p^2 - k^2)\right)\right|}\right) \quad \text{as } N \to \infty.$$

$$(6.96)$$

Let us estimate the expression

$$\min\limits_{\substack{(p,c)\neq(k,s) \\ p,c,k,s \leq N}} \left|\frac{\alpha\pi^2}{(l_1 l_2)^2}\left(l_1^2(c^2 - s^2) + l_2^2(p^2 - k^2)\right)\right|.$$

Suppose that $\chi = \left(\frac{l_2}{l_1}\right)^2 > 0,$

$$1 - N^2 \leq c^2 - s^2 = (c - s)(c + s) = mq \leq N^2 - 1,$$
$$1 - N^2 \leq p^2 - k^2 = (p - k)(p + k) = m'q' \leq N^2 - 1.$$

Then we have

$$\min\limits_{\substack{(p,c)\neq(k,s) \\ p,c,k,s \leq N}} \left|\frac{\alpha\pi^2}{(l_1 l_2)^2}\left(l_1^2(c^2 - s^2) + l_2^2(p^2 - k^2)\right)\right| = \frac{\alpha\pi^2}{l_2^2} \min\limits_{\substack{|mq| \leq N^2-1, \\ |m'q'| \leq N^2-1}} |mq + \chi m'q'|.$$

If χ is an irrational algebraic number of order $n^* \geq 2$, then Liouville's theorem implies that

$$|mq + \chi m'q'| = |m'q'||\chi + \frac{mq}{m'q'}| > \frac{C|m'q'|}{|m'q'|^{n^*}} = \frac{C}{|m'q'|^{n^*-1}},$$

where C is a positive constant depending on χ only. If

$$1 \leq c \leq N, \quad 1 \leq s \leq N, \quad 1 \leq p \leq N, \quad 1 \leq k \leq N,$$

then

$$\inf_{\substack{(m,m')\neq(0,0), \\ 2\leq q\leq 2N, \\ 2\leq q'\leq 2N}} |mq + \chi m'q'| > \inf \frac{C}{|m'q'|^{n^*-1}} = \frac{C}{\sup |m'q'|^{n^*-1}}$$

$$= \frac{C}{(2N(N-1))^{n^*-1}}.$$

Therefore,

$$\min_{\substack{(p,c)\neq(k,s) \\ p,c,k,s\leq N}} \left| \frac{\alpha\pi^2}{(l_1 l_2)^2} \left(l_1^2(c^2 - s^2) + l_2^2(p^2 - k^2) \right) \right| = \frac{\alpha\pi^2}{l_2^2} \frac{C}{(2N(N-1))^{n^*-1}}.$$

$$(6.97)$$

By substituting (6.97) into (6.96), we obtain:

$$r_N = O\left(\tau N^{2n^*-1} \sum_{p,c=1}^{N} (pc)^{2-2\gamma}(c^\gamma + p^\gamma)^2 \sum_{p,c=N+1}^{\infty} (pc)^{-2} \right) \quad \text{as } N \to \infty.$$

$$(6.98)$$

To estimate the sums in (6.98), we apply the integral test:

$$\sum_{p,c=1}^{N} (pc)^{2-2\gamma}(c^\gamma + p^\gamma)^2 \leq \int_0^N p^{2-2\gamma}dp \int_1^{N+1} c^2 dc + 2\int_0^N p^{2-\gamma}dp \int_0^N c^{2-\gamma}dc$$

$$+ \int_1^{N+1} p^2 dp \int_0^N c^{2-2\gamma}dc = \frac{2N^{6-2\gamma}}{(3-\gamma)^2} + \frac{2N^{3-2\gamma}((N+1)^3 - 1)}{9 - 6\gamma},$$

and, for $\gamma \neq 3$ and $\gamma \neq \frac{3}{2}$, we conclude that

$$\sum_{p,c=N+1}^{\infty} (pc)^{-2} \leq \int_N^{\infty} dp \int_N^{\infty} (pc)^{-2}dc = \int_N^{\infty} p^{-2}dp \int_N^{\infty} c^{-2}dc = \frac{1}{N^2}.$$

Let us substitute these integrals into (6.98) to obtain the following estimate:

$$r_N = O\left(\frac{\tau 2 N^{2n^*-1} N^{6-2\gamma}}{N^2(3-\gamma)^2} + \frac{\tau 2 N^{2n^*-1} N^{3-2\gamma}((N+1)^3-1)}{N^2(9-6\gamma)}\right).$$

The above formula can be equivalently written as

$$r_N = O\left(\tau N^{2n^*-2\gamma+3}\right). \tag{6.99}$$

This representation holds for all values $\tau = O(N^{n^*-1})$ satisfying inequality (6.93). To complete the proof, we use a value of γ from condition (3) of this Theorem:

$$\gamma > \frac{3}{2}n^* + 1.$$

Then $r_N \to 0$ as $N \to \infty$ in estimate (6.99). Therefore, property (6.86) holds for sufficiently large N.

The assertion of Theorem 6.6 can be reformulated in terms of the closure of the reachable set as follows.

Corollary 6.1 *Let the conditions of Theorem 6.6 be satisfied. Then* $q^1 \in \overline{R(0)}$.

Remark 6.2 Note that the reachable set at some time $T_0 > 0$ has been estimated in the monograph [3, Chap. VIII] for a beam-plate system with vanishing initial data when the control acts on a part of the boundary of the plate.

In this chapter, we have considered the control as the angular acceleration of the carrier body.

References

1. Lagnese, J.E.: Boundary Stabilization of Thin Plates. SIAM, Philadelphia (1989)
2. Lagnese, J., Lions, J.L.: Modelling Analysis and Control of Thin Plates. Springer, Berlin (1988)
3. Lagnese, J.E., Leugering, G., Schmidt, E.J.P.G.: Modeling, Analysis and Control of Dynamic Elastic Multi-Link Structures. Springer, New York (1994)
4. Lasiecka, I.: Mathematical Control Theory of Coupled PDEs. SIAM, Philadelphia (2002)
5. Ammari, K., Nicaise, S.: Stabilization of a transmission wave/plate equation. J. Differ. Equ. **249**, 707–727 (2010)
6. Bradley, M.E., Lenhart, S.: Bilinear spatial control of the velocity term in a Kirchhoff plate equation. Electron. J. Diff. Equ. **2001**(27), 1–15 (2001)
7. Gorain, G.C.: Boundary stabilization of nonlinear vibrations of a flexible structure in a bounded domain in R^n. J. Math. Anal. Appl. **319**, 635–650 (2006)
8. Guo, Y., Yao, P.: Stabilization of Euler-Bernoulli plate equation with variable coefficients by nonlinear boundary feedback. J. Math. Anal. Appl. **317**, 50–70 (2006)
9. Park, J.Y., Kang, Y.H., Park, S.H.: Adaptive stabilization of a von Karman plate equation with a boundary output feedback control. Comput. Math. Appl. **58**, 1742–1754 (2009)

Printed by Printforce, the Netherlands